RevisionGuide

GCSEPhysics

Collins · do brilliantly

RevisionGuide

GCSEPhysics

Malcolm Bradley
and Chris Sunley

Series editor: Jayne de Courcy

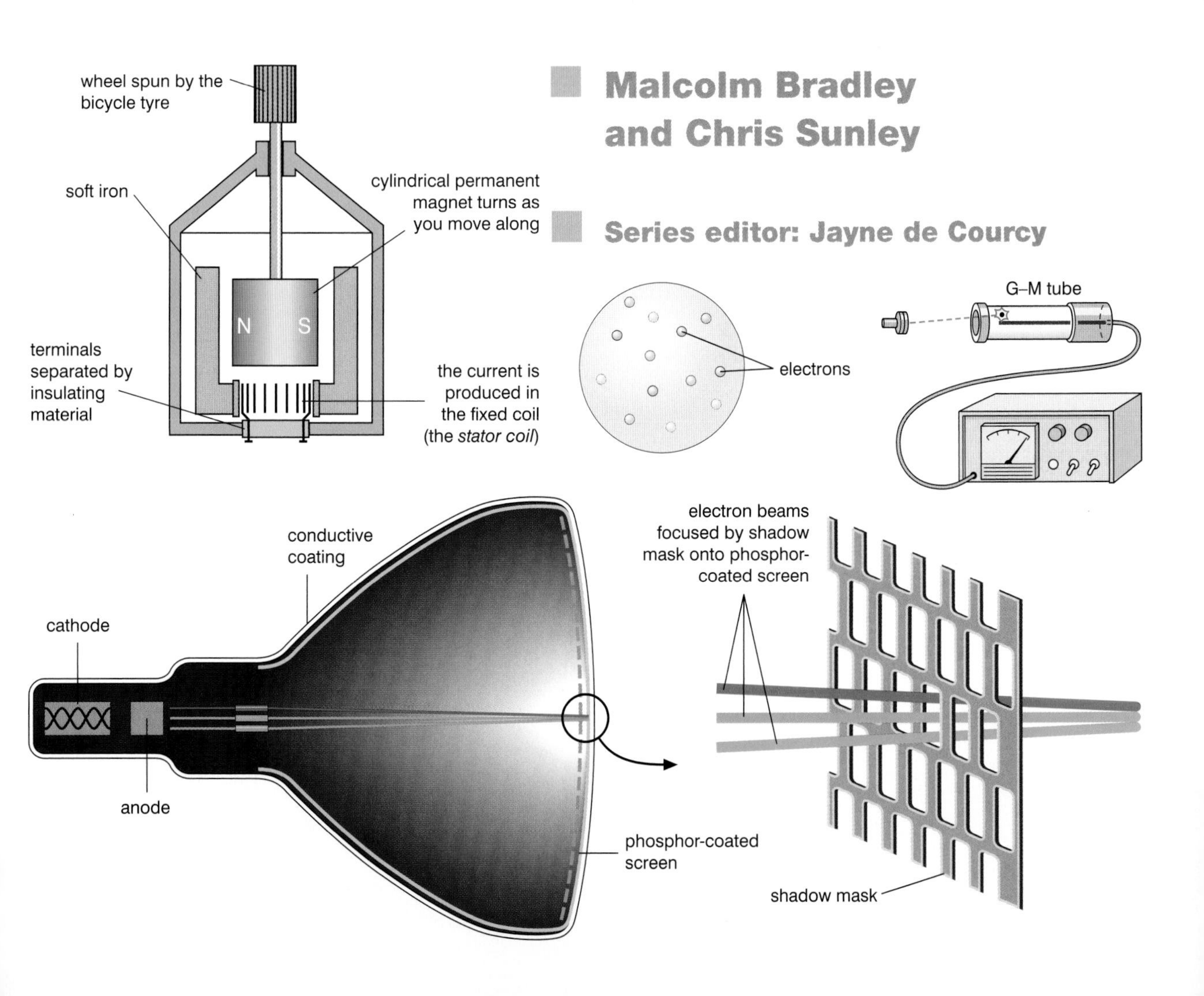

CONTENTS AND REVISION PLANNER

About this book

Exams are about much more than just repeating memorised facts, so we have planned this book to make your revision as active and effective as possible.

How?

- by breaking down the content into manageable chunks (Revision Sessions)
- by testing your understanding at every step of the way (Check Yourself Questions)
- by providing extra information to help you aim for the very top grade (A* Extras)
- by highlighting Ideas and Evidence topics (Ideas and Evidence)
- by listing the most likely exam questions for each topic (Question Spotters)
- by giving you invaluable examiner's guidance about exam technique (Exam Practice)

REVISION SESSION I

Revision Sessions

- Each Unit is divided into a number of **short revision sessions**. You should be able to read through each of these in no more than 30 minutes. That is the maximum amount of time that you should spend on revising without taking a short break.
- Ask your teacher for a copy of your own exam board's **GCSE Physics specification**. Tick off on the Contents list each of the revision sessions that you need to cover. It will probably be most of them.

? Check Yourself Questions

- At the end of each revision session there are some **Check Yourself Questions**. By trying these questions, you will immediately find out whether you have understood and remembered what you have read in the revision session. **Answers** are at the back of the book, along with **extra hints and guidance**.
- If you manage to give correct answers to all the Check Yourself questions for a session, then you can confidently tick off this topic in the box headed '**Revised & understood**'. If not, you will need to tick the '**Revise again**' box to remind yourself to return to this topic later in your revision programme.

A* EXTRA

These boxes occur in most revision sessions. They contain some **extra information** which you need to learn if you are aiming to achieve the very top grade. If you have the chance to use these additional facts in your exam, it could make the difference between a good answer and a very good answer.

IDEAS AND EVIDENCE

Physics GCSE specifications have particular topics highlighted as 'ideas and evidence'. **Every Foundation and Higher Tier paper** must have a question on one of these topics.

The boxes in the book give you guidance on the sorts of ideas, applications and social, economic and environmental issues you may be asked about in your exam.

It's obviously important to revise the facts, but it's also helpful to know how you might need to use this information in your exam.

The authors, who have been involved with examining for many years, know the sorts of questions that are most likely to be asked on each topic. They have put together these Question Spotter boxes so that they can help you to **focus your revision**.

Exam Practice

- This Unit gives you **invaluable guidance on how to answer exam questions well**.
- It contains some sample students' answers to typical exam questions, followed by examiner's comments on them, showing where the students gained and lost marks. Reading through these will help you get a very clear idea of what you need to do in order to score **full marks** when answering questions in your GCSE Physics exam.
- There are also some **typical exam questions** for you to try answering. Model answers are given at the back of the book for you to check your own answers against. There are also examiner's comments, highlighting **how to achieve full marks**.

About your GCSE Physics course

What is covered in my course?

GCSE Physics courses are all based on the National Curriculum Programme of Study for Key Stage 4. This lists the Physics that is included in all double award Science courses.

If you are studying for **GCSE Physics as a separate subject**, then your course includes all the Physics material from the double award specification and some additional topics. The examining groups have not all chosen the same topics, so it is **vital** that you make sure you know which topics apply to you.

To make sure you cover the correct topics:

- Use the matching grid opposite as a guide.
- Refer to the exam board **specification** for your course.
- Check with your teachers that you have identified the correct areas.

Foundation and Higher Tier papers

You will be entered for the exam at one of two **tiers** – either Foundation or Higher. This table shows which **grades** are available for each tier.

Higher Tier							
A*	A	B	C	D	E	F	G
			Foundation Tier				

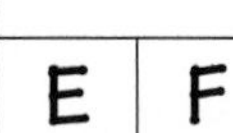

Check the specification for your course – it will tell you exactly which material is needed for each tier.

Answering the questions

When you are deciding how to answer an examination question, remember these key points:

- How many marks are available?
  You will need to give at least as many points in your answer as there are marks available.
- How much space is there for the answer?
  The number of lines available for your answer is usually a good idea of the detail the examiner is expecting in your response.
- What type of question is it?
  Recognising the type of question can help you give the correct level of response.

The main types of question

- **Multiple choice questions**
  These can take several forms – choosing the correct letter from a list of answers, putting a ring around the correct word in a list and so on. In each case no further detail is required (unless the follow-up question is 'explain your answer') and the examiner is checking your knowledge of straightforward ideas such as vocabulary, definitions and units.
- **Short answer questions**
  These are very common on GCSE question papers. They are usually worth one or two marks and have one or two lines for the answer. Often, only a few words are needed to get across the essential information. Check the mark schemes provided in this book and by the exam boards to get a feel for this kind of question.
- **Extended response questions**
  These questions are typically worth three or four marks. The examiner is looking to see how you can connect ideas together to form a logical sequence. For this type of question, take your time and think the sequence through carefully.

Quality of written communication

Some marks on your exam will cover this area. Remember to use capital letters at the start of sentences and full stops at the end. Check your spelling carefully and try to use the correct technical words where you can.

Ideas and evidence

How scientific ideas grow, and how they are communicated, will form part of your exam. Often this will mean that quite a lot of information is given in the question. Don't panic over questions like this – you may not be familiar with the particular example being used, but the whole point is that a number of themes apply to scientific development as a whole. Use the 'Ideas and evidence' boxes in this book to see examples of these themes.

MATCHING GRID

This grid gives an indication as to which revision sessions apply to the different exam board specifications.

If you are studying GCSE Science double award, then you need to look at the first column in the table.

If you are studying GCSE Physics, then you need to find the correct column for your specification as well as the double award column – remember that GCSE Physics includes all the Double Award material as well as some extra material specific to your exam board.

For each revision session, if there is a • in the grid then you need to study the whole of that revision session. If there is a • in the grid then you need to study part of the material in that revision session – you should check with your teacher exactly which parts are needed.

			Double Award	AQA	Edexcel	OCR option A	OCR option B	WJEC
Unit 1 Electricity								
Revision session	1	Electric circuits	•					
	2	What affects resistance?	•					
	3	Power in electrical circuits	•					
	4	Static electricity	•					
Unit 2 Electromagnetic effects								
Revision session	1	Electromagnetism	•					
	2	Electromagnetic induction	•					
	3	More electromagnetic devices			•		•	
Unit 3 Electronics and control								
Revision session	1	Logic gates		•		•		•
	2	Input sensors and output devices		•		•		•
	3	Electronic systems		•		•		•
Unit 4 Forces and motion								
Revision session	1	The effects of forces	•				•	•
	2	Velocity and acceleration	•			•	•	•
	3	Vehicle safety features				•		
	4	Using the equations of motion				•	•	•
	5	Motion in two dimensions				•	•	•
	6	Momentum		•		•	•	•
	7	Turning forces						•
Unit 5 Energy								
Revision session	1	Where does our energy come from?	•					
	2	Transferring energy	•					
	3	Specific heat capacity				•	•	•
	4	Work, power and energy	•					
Unit 6 Describing waves								
Revision session	1	The properties of waves	•					
	2	The electromagnetic spectrum	•					
	3	Light reflection and refraction	•	•		•		•
	4	Sound waves, resonance and musical instruments	•			•	•	
	5	Interference				•		•
Unit 7 Waves for communication								
Revision session	1	Information transfer and storage			•		•	
	2	Radio systems			•		•	
Unit 8 The Earth and beyond								
Revision session	1	The Solar System	•	•	•			
	2	How did the Universe begin?	•					
Unit 9 Particles								
Revision session	1	Kinetic theory of gases			•		•	
	2	Developing ideas about atoms	•		•			•
	3	Unstable atoms	•					
	4	Uses and dangers of radioactivity	•					
	5	Fundamental particles			•			
	6	Electron beams			•		•	•

REVISION SESSION 1 Electric circuits

What does electricity do?

- You cannot see electricity but you can see the effects it has. It is very good at **transferring energy**, and can:
  - make things **hot** – as in the heating element of an electric fire
  - make things **magnetic** – as in an electromagnet
  - produce **light** – as in a light bulb
  - **break down** certain compounds and solutions – as in electrolysis.

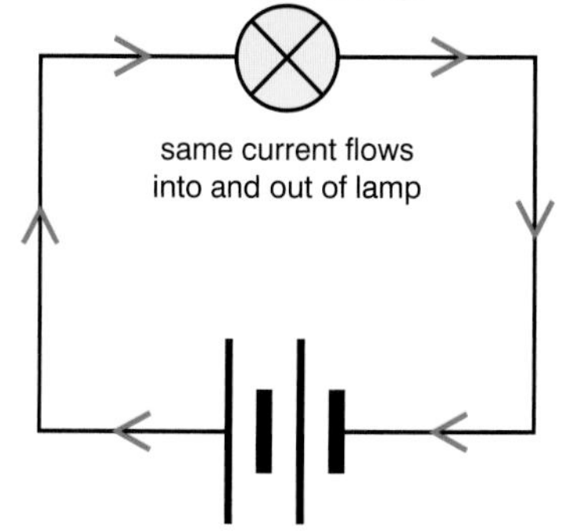

In this simple circuit the arrows show the direction of the current.

A simple model of an electrical circuit

- The battery in an electrical circuit can be thought of as pushing electrical charge round the circuit to make a current. It also transfers energy to the electrical charge. The **voltage** of the battery is a measure of how much 'push' it can provide and how much energy it can transfer to the charge.
- Scientists now know that electric current is really a **flow of electrons** around the circuit from negative to positive. Unfortunately, early scientists guessed the direction of flow incorrectly. Consequently all diagrams were drawn showing the current flowing from positive to negative. This way of showing the current has not been changed and so the **conventional current** that everyone uses gives the direction that positive charges would flow.

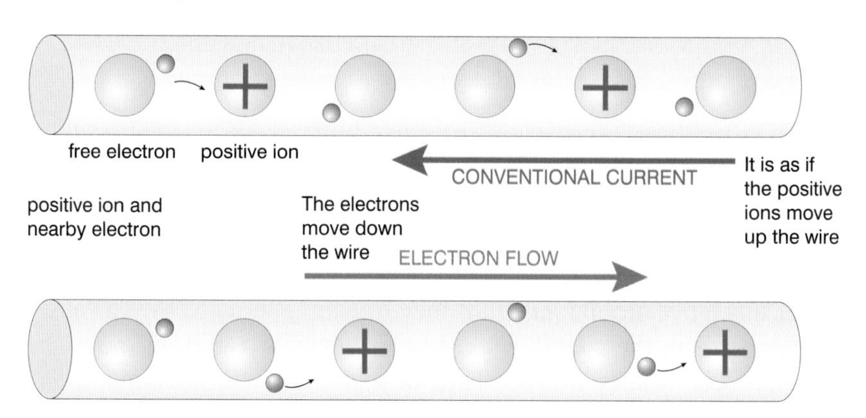

Conventional current is drawn in the opposite direction to electron flow.

- There are two different ways of connecting two lamps to the same battery. Two very different kinds of circuit can be made. These circuits are called **series** and **parallel** circuits.

	Series	Parallel
Circuit diagram		
Appearance of the lamps	Both lamps have the same brightness, both lamps are dim.	Both lamps have the same brightness, both lamps are bright.
Battery	The battery is having a hard time pushing the same charge first through one bulb, then another. This means less charge flows each second, so there is a low current and energy is slowly transferred from the battery.	The battery pushes the charge along two alternative paths. This means more charge can flow around the circuit each second, so energy is quickly transferred from the battery.
Switches	The lamps cannot be switched on and off independently.	The lamps can be switched on and off independently by putting switches in the parallel branches.
Advantages/ disadvantages	A very simple circuit to make. The battery will last longer. If one lamp 'blows' then the circuit is broken so the other one goes out too.	The battery will not last as long. If one lamp 'blows' the other one will keep working.
Examples	Christmas tree lights are often connected in series.	Electric lights in the home are connected in parallel.

Charge, current and potential difference

- Electric charge is measured in **coulombs** (C). Electric current is measured in **amperes** (A).
- The electric current is the amount of charge flowing every second – the number of coulombs per second:

$$I = \frac{Q}{t}$$

I = current in amperes (A)
Q = charge in coulombs (C)
t = time in seconds (s)

- The electrons moving round a circuit have some **potential energy**. As electrons move around a circuit, they transfer energy to the various components in the circuit. For example, when the electrons move through a lamp they transfer some of their energy to the lamp.
- The amount of energy that a unit of charge (a coulomb) transfers between one point and another (the number of joules per coulomb) is called the **potential difference** (p.d.). Potential difference is measured in **volts** and so it is often referred to as **voltage**:

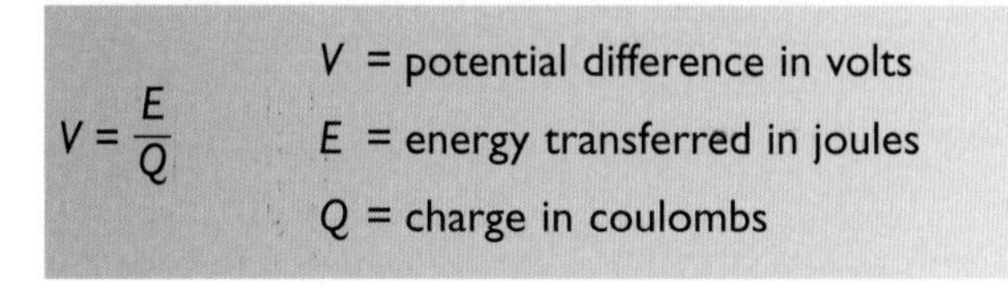

$$V = \frac{E}{Q}$$

V = potential difference in volts
E = energy transferred in joules
Q = charge in coulombs

A* EXTRA

- In a conductor, the charge carriers are electrons and flow round the circuit from negative to positive.
- In an electrolyte (e.g. sulphuric acid in a car battery), the charge carriers are ions. The negative ions (anions) move to the anode and the positive ions (cations) move to the cathode.

QUESTION SPOTTER

- You will be expected to remember the equation that links current, charge and time and carry out calculations which use it. Don't forget that charge is measured in coulombs and current in amps.

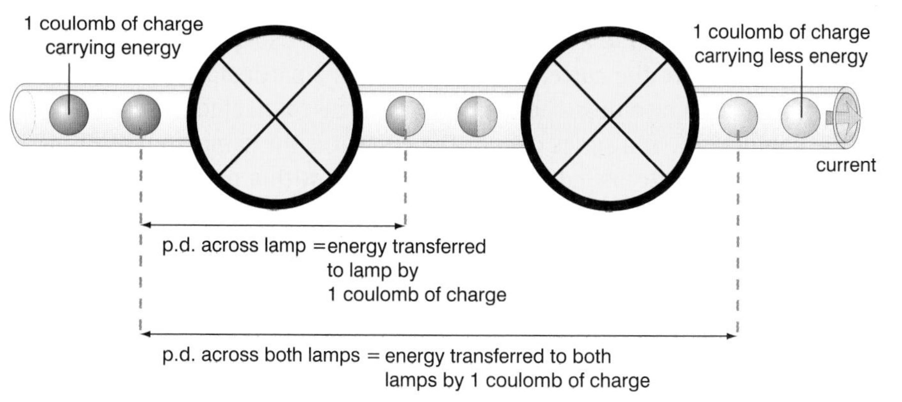

Potential difference (p.d.) is the difference in energy of a coulomb of charge between two parts of a circuit.

A* EXTRA

- The potential difference is measured between two points in a circuit. It is like an electrical pressure difference and measures the energy transferred per unit of charge flowing.

Measuring electricity

- Potential difference is measured using a **voltmeter.** If you want to measure the p.d. across a component then the voltmeter must be connected **in parallel** to that component. Testing with a voltmeter does not interfere with the circuit.
- A voltmeter can be used to show how the potential difference varies in different parts of a circuit. In a series circuit you find different values of the voltage depending on where you attach the voltmeter. You can assume that energy is only transferred when the current passes through electrical components such as lamps and motors – the energy transfer as the current flows through copper connecting wire is very small. It is only possible therefore to measure a p.d. or voltage across a component.

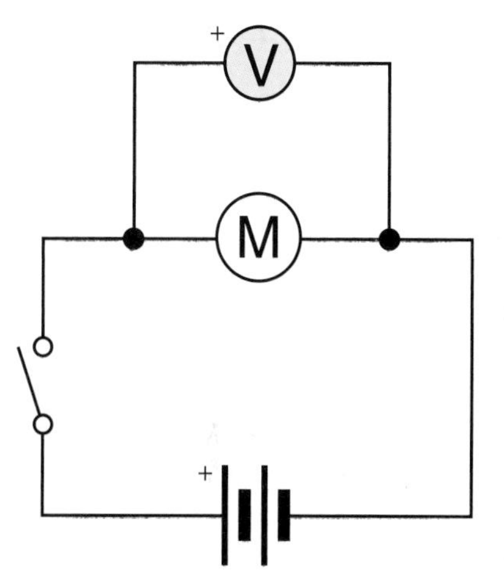

The voltmeter can be added after the circuit has been made.

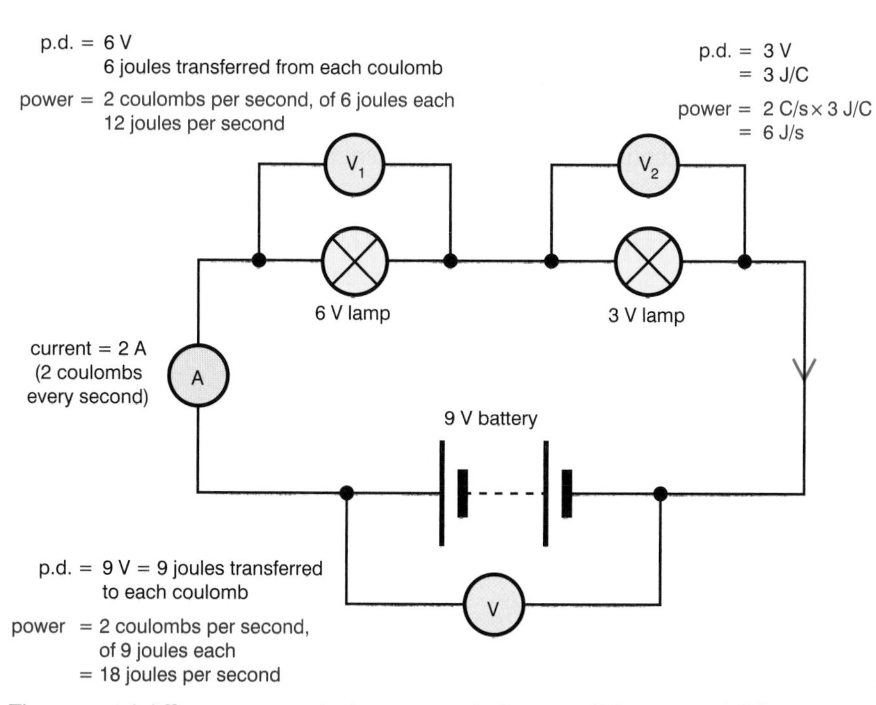

The potential difference across the battery equals the sum of the potential differences across each lamp. That is $V = V_1 + V_2$.

A* EXTRA

- The current is the same in all parts of a series circuit but the potential difference across different components can be different.
- Remember that the p.d. across the cell or battery will be equal to the sum of the p.d.s across all the components in the circuit.

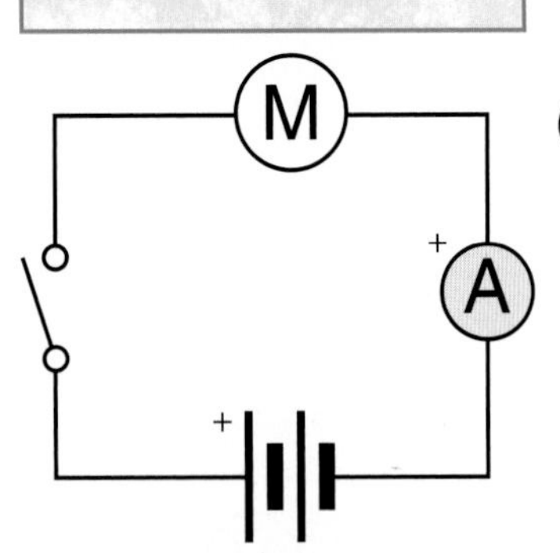

The circuit has to be broken to include the ammeter.

- The current flowing in a circuit can be measured using an **ammeter**. If you want to measure the current flowing through a particular component, such as a lamp or motor, the ammeter must be connected **in series** with the component. In a series circuit, the current is the same no matter where the ammeter is put. This is not the case with a parallel circuit.

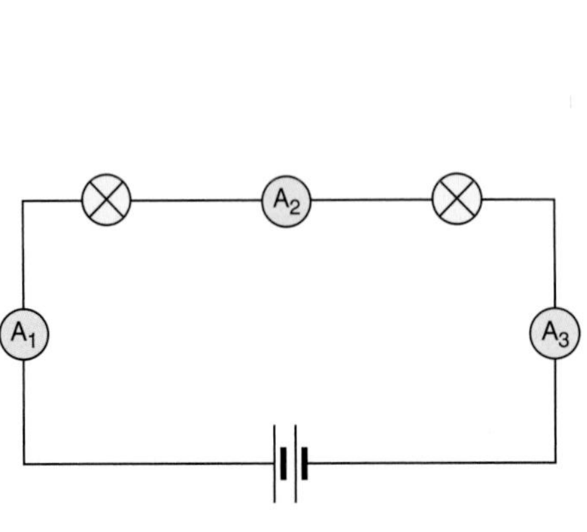

In this series circuit, the current will be the same throughout the circuit so $A_1 = A_2 = A_3$.

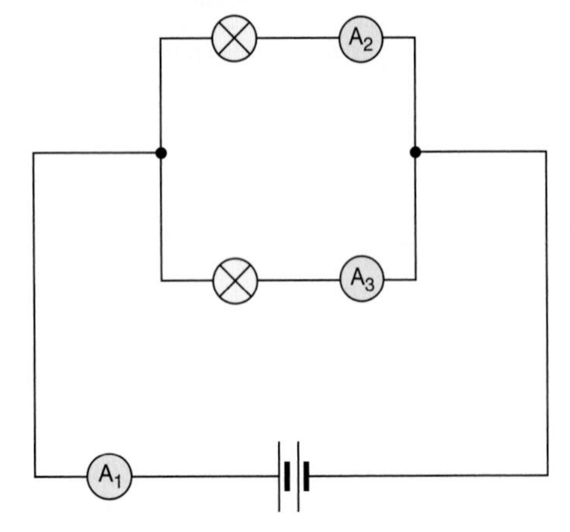

The current flow splits between the two branches of the parallel circuit so $A_1 = A_2 + A_3$.

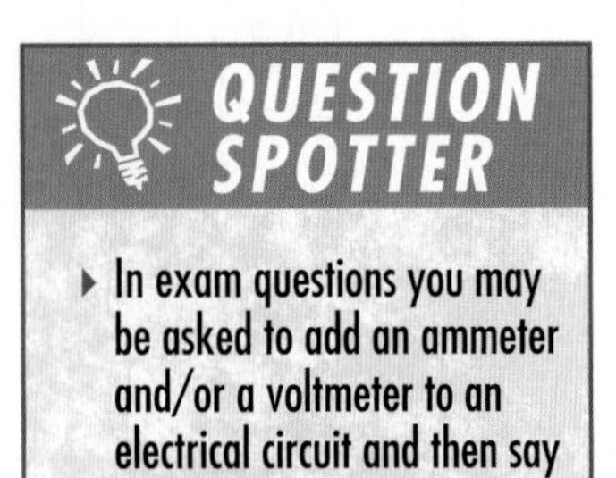

QUESTION SPOTTER

- In exam questions you may be asked to add an ammeter and/or a voltmeter to an electrical circuit and then say what readings you would expect.

Direct and alternating currents

- A battery produces a steady current. The electrons are constantly flowing from the negative terminal of the battery round the circuit and back to the positive terminal. This produces a **direct current** (d.c.).
- The mains electricity used in the home is quite different. The electrons in the circuit move backwards and forwards. This kind of current is called **alternating current** (a.c.). Mains electricity moves forwards and backwards 50 times each second, that is, with a frequency of 50 hertz (Hz).
- The advantage of using an a.c. source of electricity rather than a d.c. source is that it can be transmitted from power stations to the home at very high voltages, which reduces the amount of energy that is lost in the overhead cables (see Unit 2).

IDEAS AND EVIDENCE

Electrical ideas were developed during the 1800s before the discovery of the electron. This led to some mistakes, such as showing the current moving the 'wrong' way.

CHECK YOURSELF QUESTIONS

Q1 Look at the following circuit diagrams. They show a number of ammeters and in some cases the readings on these ammeters. All the lamps are identical.

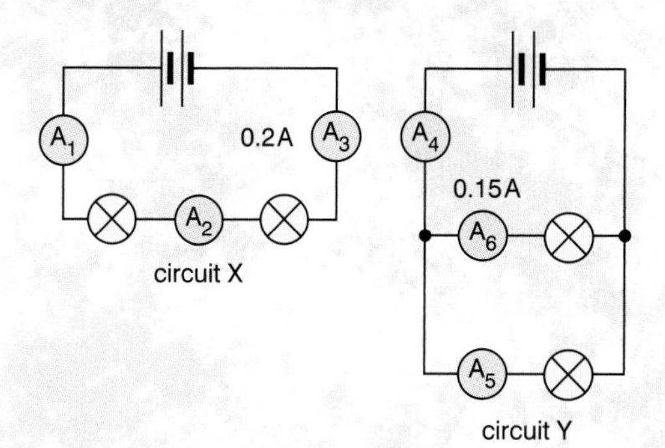

a For circuit X, what readings would you expect on ammeters A_1 and A_2?

b For circuit Y, what readings would you expect on ammeters A_4 and A_5?

Q2 Look at the circuit diagram. It shows how three voltmeters have been added to the circuit. What reading would you expect on V_1?

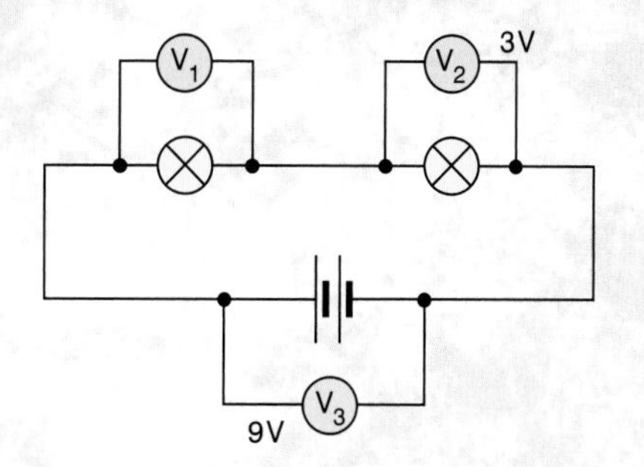

Q3 **a** A charge of 10 coulombs flows through a motor in 30 seconds. What is the current flowing through the motor?

b A heater uses a current of 10 A. How much charge flows through the lamp in: **i** 1 second, **ii** 1 hour?

Answers are on page 135.

What affects resistance?

Ohm's law

- The relationship between voltage, current and resistance in electrical circuits is given by **Ohm's law**:

	V is the voltage in volts (V)
$V = I R$	I is the current in amps (A)
	R is the resistance in ohms (Ω)

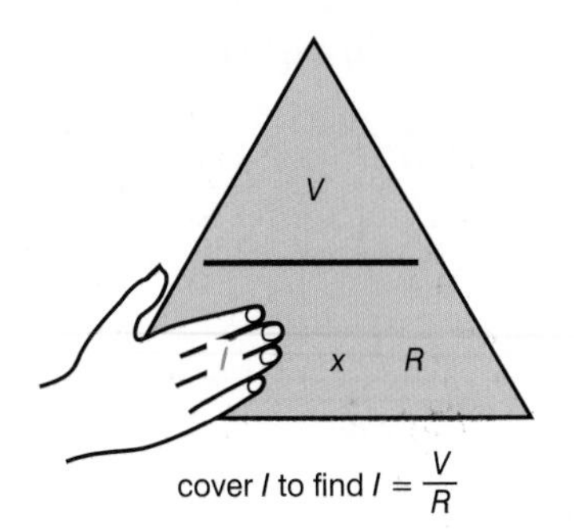

cover I to find $I = \frac{V}{R}$

- It is important to be able to rearrange this equation when performing calculations. Use the triangle on the left to help you.

WORKED EXAMPLES

1 Calculate the resistance of a heater element if the current is 10 A when it is connected to a 230 V supply.

Write down the formula in terms of R:	$R = \frac{V}{I}$
Substitute the values for V and I:	$R = \frac{230}{10}$
Work out the answer and write down the unit:	$R = 23\ \Omega$

2 A 6 V supply is applied to 1000 Ω resistor. What current will flow?

Write down the formula in terms of I:	$I = \frac{V}{R}$
Substitute the values for V and R:	$I = \frac{6}{1000}$
Work out the answer and write down the unit:	$I = 0.006$ A

QUESTION SPOTTER

- Calculations involving Ohm's law are very common. You will need to remember the equation and be able to change the subject. Marks will usually be given for the correct units.

Effect of material on resistance

- Substances that allow an electric current to flow through them are called **conductors**; those which do not are called **insulators**.
- Metals are conductors. In a metal structure, the metal atoms exist as ions surrounded by an electron cloud. If a potential difference is applied to the metal, the electrons in this cloud are able to move and a current flows.

In a metal structure metal ions are surrounded by a cloud or 'sea' of electrons.

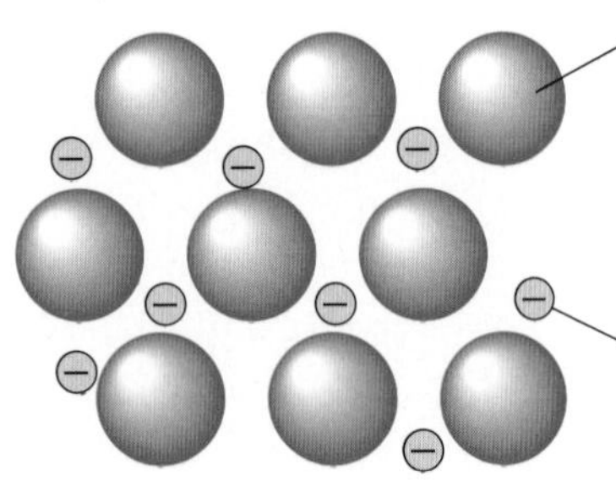

metal atoms (some people describe them as positive ions because they donate electrons into the 'sea' of electrons)

'sea' of electrons holds the metal atoms together

- When the electrons are moving through the metal structure, they bump into the metal ions and this causes **resistance** to the electron flow or current. In different conductors the ease of flow of the electrons is different and so the conductors have different resistances. For instance, copper is a better conductor than iron.

Effects of length and cross-sectional area

- For a particular conductor, the resistance is **proportional to length**. The longer the conductor, the further the electrons have to travel, the more likely they are to collide with the metal ions and so the greater the resistance.

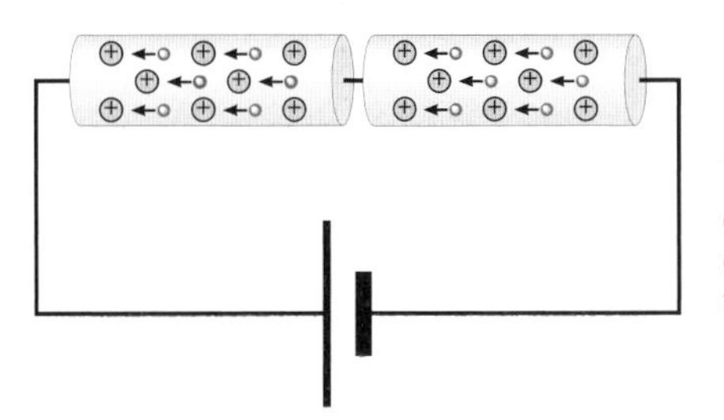

Two wires in series are like one long wire, because the electrons have to travel twice as far.

- Resistance is **inversely proportional to cross-sectional area.** The greater the cross-sectional area of the conductor, the more electrons there are available to carry the charge along the conductor's length and so the lower the resistance.

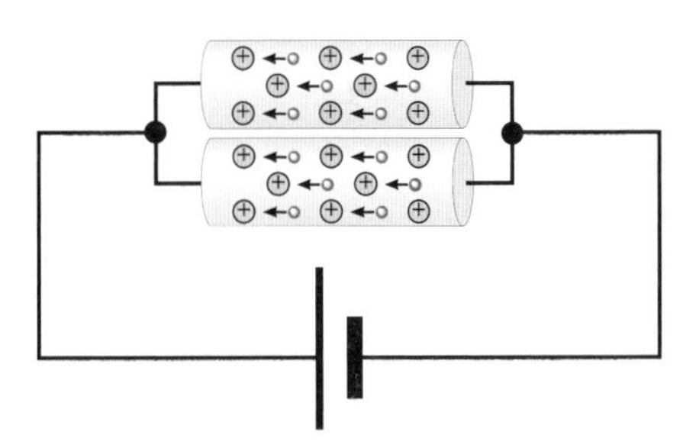

Two wires in parallel are like one thick wire, so the electrons have more routes to travel along the same distance.

- The amount of current flowing through a circuit can be controlled by changing the resistance of the circuit using a **variable resistor** or **rheostat**. Adjustment of the rheostat changes the length of the wire the current has to flow through.

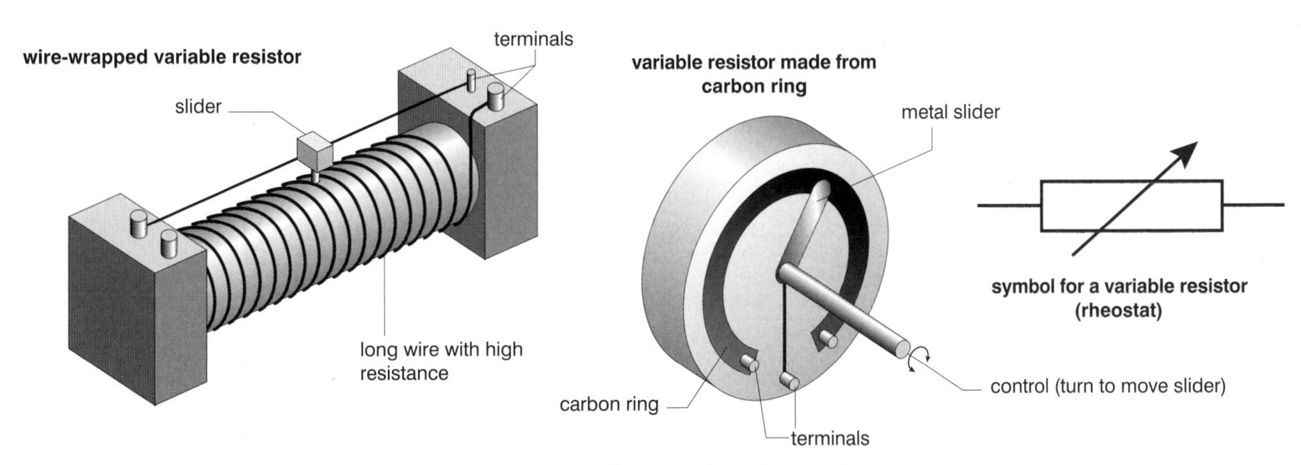

Variable resistors are commonly used in electrical equipment, for example in the speed controls of model racing cars or in volume controls on radios and hi-fi systems.

QUESTION SPOTTER

- Questions often ask you to state and explain how the resistance of a conductor is affected by its length or its thickness. Using the simple models shown here should help.

IDEAS AND EVIDENCE

Using the idea of electrons to explain how a material behaves is a good example of scientists using a mental model.

Effect of temperature on resistance

- If the resistance of a conductor remains constant, a graph of voltage against current will give a **straight line**. The gradient of the line will be the resistance of the conductor.
- The resistance of most conductors becomes higher if the temperature of the conductor increases. As the temperature rises, the metal ions vibrate more and provide greater resistance to the flow of the electrons. For example, the resistance of a filament lamp becomes greater as the voltage is increased and the lamp gets hotter.

In an 'ohmic' resistor, such as carbon, Ohm's law applies and the voltage is directly proportional to the current – a straight line is obtained. In a filament lamp, Ohm's law is not obeyed because the heating of the lamp changes its resistance.

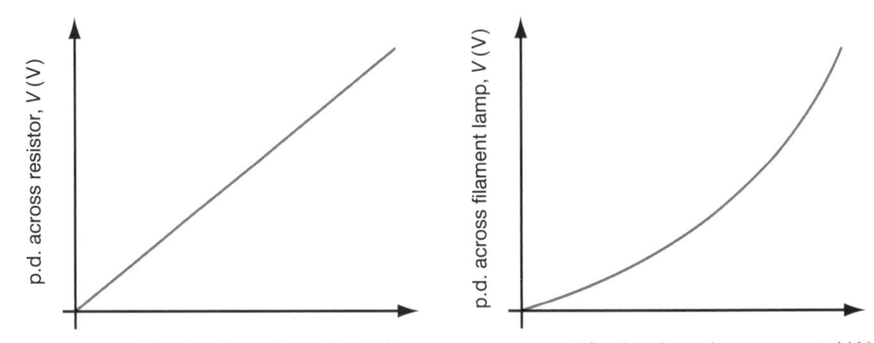

- You will often be shown graphs of potential difference against current and asked to explain how the current varies with the p.d.

A* EXTRA

- In a thermistor, increasing temperature actually reduces the resistance. This is the opposite effect to that in a normal resistor.
- In a light-dependent resistor (LDR) an increase in brightness reduces the resistance.

- In some substances, increasing the temperature actually **lowers** the resistance. This is the case with **semiconductors** such as silicon. Silicon has few free electrons and so behaves more like an insulator than a conductor. But if silicon is heated, more electrons are removed from the outer electron shells of the atoms producing an increased electron cloud. The released electrons can move throughout the structure, creating an electric current. This effect is large enough to outweigh the increase in resistance that might be expected from the increased movement of the silicon ions in the structure as the temperature increases.
- Semiconducting silicon is used to make **thermistors**, which are used as temperature sensors, and **light-dependent resistors** (LDRs), which are used as light sensors.

A light-dependent resistor, (top left) conducts better when light shines on it. A thermistor (top right) conducts better when it is hot. A diode (bottom left) only conducts in one direction. An ordinary (ohmic) resistor is shown bottom right.

- In LDRs it is light energy that removes electrons from the silicon atoms, increasing the electron cloud.

- Silicon **diodes** also use resistance as a means of controlling current flow in a circuit. In one direction, the resistance is very high and effectively prevents current flow. In the other direction, the resistance is relatively low and current can flow. Diodes are used to **protect** sensitive electronic equipment that would be damaged if a current flowed in the wrong direction.

CHECK YOURSELF QUESTIONS

Q1 **a** Draw a circuit diagram to show how you could measure the resistance of a piece of nichrome wire. Explain how you would calculate the resistance of the wire.

b How would the resistance of the wire change if:

i its length was doubled,

ii its cross-sectional area was doubled?

Q2 Use Ohm's law to calculate the following:

a The voltage required to produce a current of 2 A in a 12 Ω resistor.

b The voltage required to produce a current of 0.1 A in a 200 Ω resistor.

c The current produced when a voltage of 12 V is applied to a 100 Ω resistor.

d The current produced when a voltage of 230 V is applied to a 10 Ω resistor.

e The resistance of a wire which under a potential difference of 6 V allows a current of 0.1 A to flow.

f The resistance of a heater which under a potential difference of 230 V allows a current of 10 A to flow.

Q3 A graph of current against voltage is plotted for a piece of wire. The graph is shown below.

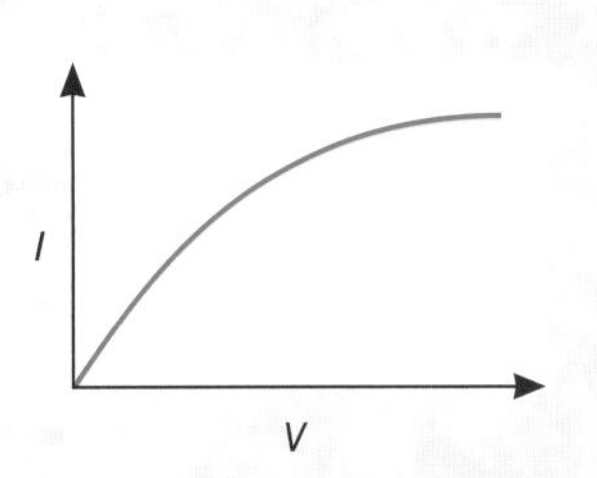

a Describe how the resistance of the wire changes as the voltage is increased.

b Suggest an explanation for this change.

Answers are on page 135.

Power in electrical circuits

Calculating power ratings

- All electrical equipment has a **power rating**, which indicates how many joules of energy are supplied each second. The unit of power used is the **watt** (W). Light bulbs often have power ratings of 60 W or 100 W. Electric kettles have ratings of about 2 kilowatts (2 kW = 2000 W). A 2 kW kettle supplies 2000 J of energy each second.
- The power of a piece of electrical equipment depends on the voltage and the current:

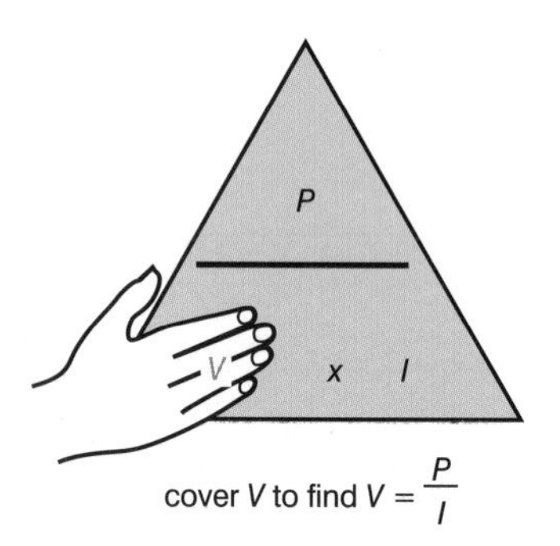

cover V to find $V = \frac{P}{I}$

$P = V\,I$	P = power in watts (W) V = voltage in volts (V) I = current in amps (A)

WORKED EXAMPLES

1 What is the power of an electric toaster if a current of 7 A is obtained from a 230 V supply?

Write down the formula in terms of P:	$P = V\,I$
Substitute the values:	$P = 230 \times 7$
Work out the answer and write down the unit:	$P = 1610$ W

QUESTION SPOTTER

- Questions on the equation $P = V\,I$ are very common. You will need to be able to change the subject of the equation and give the units for P, V and I.

2 An electric oven has a power rating of 2 kW. What current will flow when the oven is used with a 230 V supply?

Write down the formula in terms of I:	$I = \frac{P}{V}$
Substitute the values:	$I = \frac{2000}{230}$
Work out the answer and write down the unit:	$I = 8.7$ A

How do suppliers charge for electricity?

- Electricity meters in the home and in industry measure the amount of energy used in **kilowatt-hours** (kWh). 1 kilowatt-hour (1 kWh) is the amount of energy transferred by a 1 kW device in 1 hour.

  1 kWh = 3 600 000 J.

WORKED EXAMPLE

Calculate the energy transferred by a 3 kW electrical immersion heater which is used for 30 minutes.

Write down the formula:	Energy = power (kW) × time (h)
Substitute the values:	Energy = 3 × 0.5
Work out the answer and write down the unit:	Energy = 1.5 kWh

Balancing supply and demand

An off-peak white meter has two displays. The lower display shows the number of kWh units used at the standard rate. The upper display shows the units used at off-peak rate.

- Because they cannot close the power stations at night when demand is lower, the electricity companies sell electricity at two different rates.
- **Standard rate** applies to electricity used during the day.
- **Off-peak rate** electricity is used at night, and costs less than half standard rate electricity. Users need a special meter for this type of electricity. Off-peak electricity is usually available from midnight until 7 a.m.

WORKED EXAMPLE

A 3 kW night storage heater is switched on full power for 7 hours one night. The tariff is 4p per unit at night. Calculate the cost of using the heater.

Write down the formula:	Units = power (kW) × time (h)
Substitute the values:	Units = 3 × 7 = 21
Include cost of each unit:	Cost = units × 4p = 21 × 4 = 84p

Using electricity safely

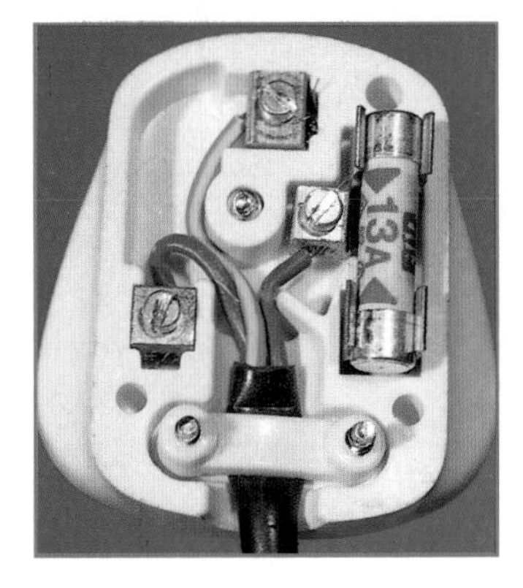

The fuse fits between the live brown wire and the pin. The brown live wire and the blue neutral wire carry the current. The green and yellow striped earth wire is needed to make metal appliances safer.

- Electrical appliances can be damaged if the current flowing through them is too high. The electric current usually has to pass through the **fuse** in the plug before it reaches the appliance. If there is a sudden surge in the current, the wire in the fuse will heat up and melt – it 'blows'. This breaks the circuit and stops any further current flowing.
- The fuse must have a value above the normal current that the appliance needs but should be as small as possible. The most common fuses are rated at 3 A, 5 A and 13 A. Any electrical appliance with a heating element in it should be fitted with a 13 A fuse.

WORKED EXAMPLES

1 What fuse should be fitted in the plug of a 2.2 kW electric kettle used with a supply voltage of 230 V?

Calculate the normal current: $I = \frac{P}{V} = \frac{2200\text{ W}}{230\text{ V}} = 9.6\text{ A}$.

Choose the fuse with the smallest rating bigger than the normal current: the fuse must be 13 A.

2 What fuse should be fitted to the plug of a reading lamp which has a 60 W lamp and a supply of 230 V?

Calculate the normal current: $I = \frac{P}{V} = \frac{60\text{ W}}{230\text{ V}} = 0.26\text{ A}$.

Choose the fuse with the smallest rating bigger than the normal current: the fuse must be 3 A.

Other safety measures

- **Circuit breakers** spring open ('trip') a switch if there is an increase in current in the circuit. They can be reset easily after the fault in the circuit has been corrected.
- Metal-cased appliances must have an **earth wire** as well as a fuse. If the live wire worked loose and came into contact with the metal casing, the casing would become live and the user could be electrocuted. The earth wire provides a very low resistance route to the 0 V earth – usually water pipes buried deep underground. This low resistance means that a large current passes from the live wire to earth, causing the fuse to melt and break the circuit.
- Appliances that are made with plastic casing do not need an earth wire. The plastic is an insulator and so can never become live. Appliances like this are said to be **double insulated**.

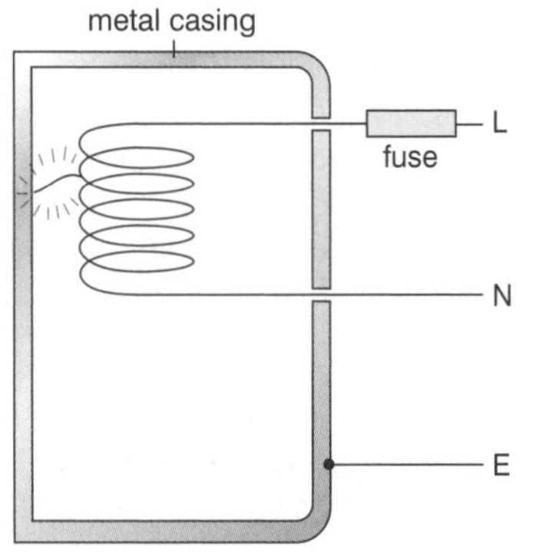

The earth wire and fuse work together to make sure that the metal outer casing of this appliance can never become live and electrocute someone.

CHECK YOURSELF QUESTIONS

Q1 A hairdryer works on mains electricity of 230 V and takes a current of 4 A. Calculate the power of the hairdryer.

Q2 **a** An iron has a power rating of 1000 W. Calculate the cost of using the iron for 3 hours if electricity costs 8p per kilowatt-hour.

b The power of a television set is 200 W. Calculate how much it costs to watch the television for 5 hours. A unit (kWh) of electricity costs 8p.

Q3 A 3 kW immersion heater transfers energy to water for 10 minutes. How much energy does the heater transfer in 10 minutes?

Answers are on page 136.

Static electricity

What is static electricity?

- Electric charge does not move about in an insulator. However, when two insulators are rubbed together, electrons from the atoms on the surface of one insulator are transferred to the surface of the other. This leaves an excess of electrons – a negative charge – on one insulator and a shortage of electrons – a positive charge – on the other insulator. The charge is fixed in place – it is **static electricity**.

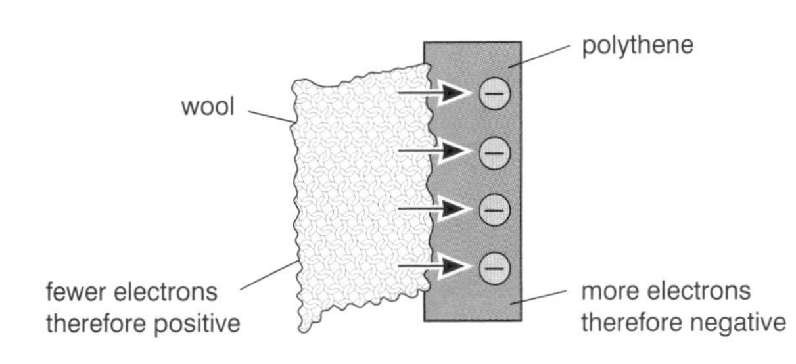

Charging up a plastic ruler.

- Static charges interact in a simple way: **like charges repel – unlike charges attract**. For example, when a balloon is rubbed against clothing it will 'stick' to a wall or ceiling. This is because of **electrostatic induction**. The balloon 'sticks' because of the attraction between the negative charges on the balloon and the induced positive charges on the ceiling.

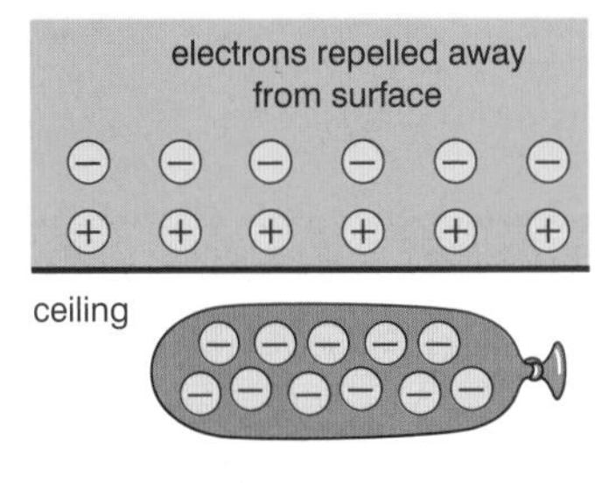

The balloon induces a charge on the ceiling's surface.

What problems can static electricity cause?

- Air currents in thunderclouds rub past water molecules and charge them up. The bottom of the cloud is left with a negative charge, which induces a positive charge in buildings and trees on the ground and creates a strong electric field in the air between the cloud and the ground. A short burst of electric current can occur in the air – **lightning**.
- The sudden discharge of electricity caused by friction between two insulators can cause **shocks in everyday situations** – for example:
  - combing your hair
  - pulling clothes over your head
  - walking on synthetic carpets
  - getting out of a car.
- Sparks can also cause **explosions** – great care must be taken when emptying fuel tankers at service stations and airports, or in grain silos or flour factories.

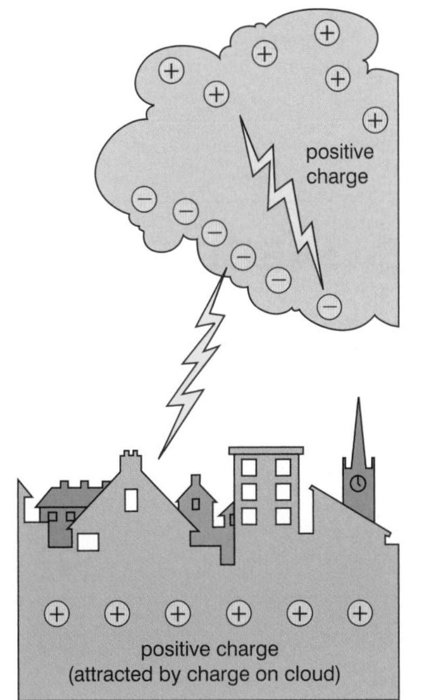

Thunderclouds have a high concentration of negative charge at the bottom. They have very high voltages. They ionise the air so that it conducts electricity. Lightning is caused by a burst of electric current through the air.

- A common question will ask you to explain how a flash of lightning occurs. It is easier to explain if you include a diagram showing the build up of charges.

- A common question will ask you to explain how static charges can be dangerous at a petrol station or when refuelling an aircraft. Don't just say 'it explodes' – you need to mention that static charges can cause sparks and that the spark can cause an explosion.

Uses of static electricity

- The properties of static electricity are put to good effect in **ink jet printers** and **photocopiers.**

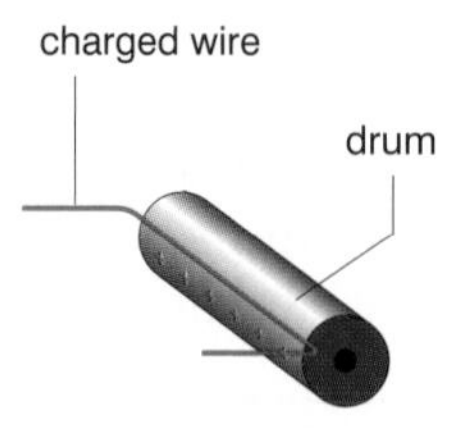

1 high voltage wire charges drum

2 drum is charged evenly

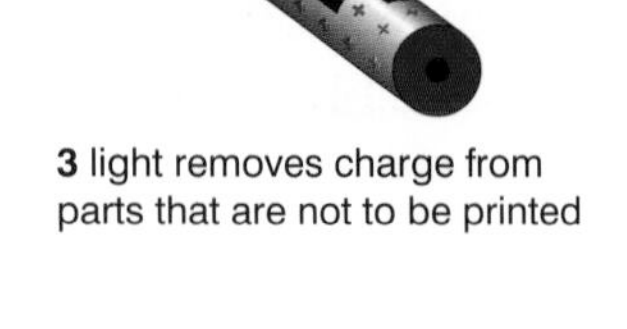

3 light removes charge from parts that are not to be printed

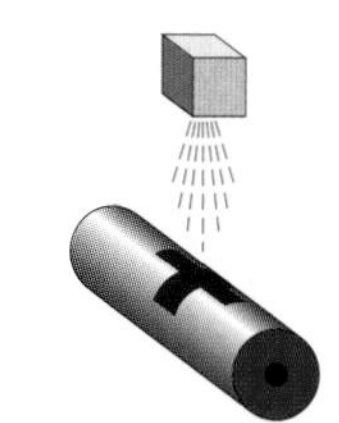

4 charged parts attract toner particles

5 toner rolled onto paper

6 toner melted on to paper by heating

In a photocopier, charged particles attract the toner. Light is used to remove charge from parts that are not to be printed.

charged ink droplets

deflection plates

ink gun

plate to stop undeflected ink

connected to earth

computer controlled power supply

In ink jet printers, uncharged ink droplets do not reach the paper. This is how the spaces between words are made.

You need to be aware of evidence supporting the idea of two 'opposite' types of charge: positive and negative. Remind yourself of the experiments involving attraction and repulsion.

CHECK YOURSELF QUESTIONS

Q1 A plastic rod is rubbed with a cloth.

a How does the plastic become positively charged?

b The charged plastic rod attracts small pieces of paper. Explain why this attraction occurs.

Q2 a A car stops and one of the passengers gets out. When she touches a metal post she feels an electric shock. Explain why she feels this shock.

b Write down two other situations where people might get this type of shock.

Q3 Give two examples of where static electricity can be dangerous.

Answers are on page 136.

REVISION SESSION 1 Electromagnetism

What is electromagnetism?

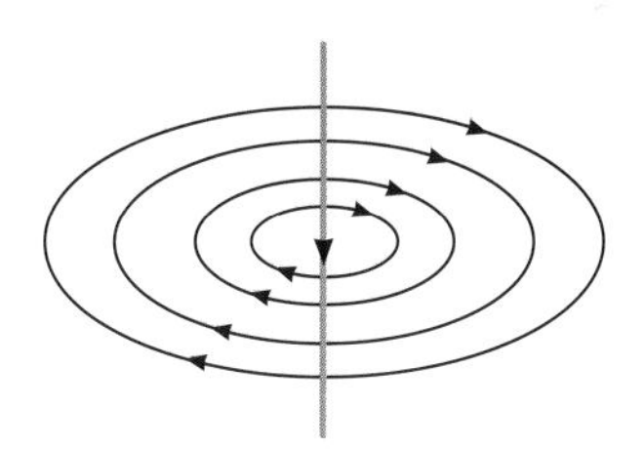

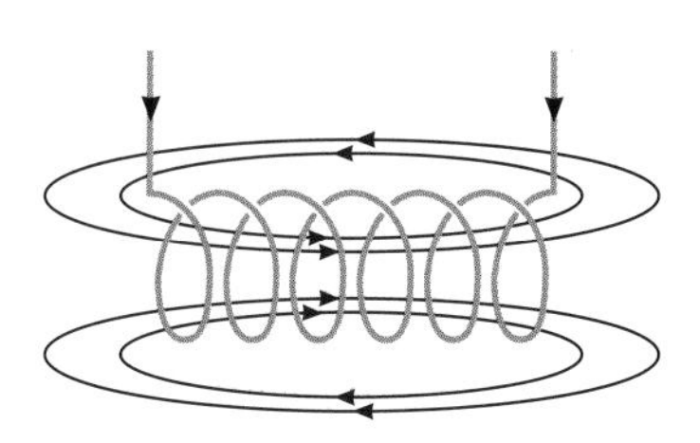

Magnetic fields are created around any wire that carries a current.

- An electric current flowing through a wire creates a magnetic field in the region of the wire. Magnetism created in this way is known as **electromagnetism**. The magnetic field is stronger if the wire is made into a coil. It is even stronger if the coil is wrapped around a piece of magnetic material such as iron.
- The strength of the electromagnet can be increased by:
  - increasing the number of coils in the wire
  - increasing the current flowing through the wire
  - placing a soft iron core inside the coils.

How do relay switches work?

- **Electromagnets** produce magnetic fields that can be turned on and off at will. A **relay** is an electromagnetic switch. It has the advantage of using a small current from a low-voltage circuit to switch on a higher current in a higher-voltage circuit.
- A relay switch operates the **starter motor** of a car. Thin low-current wires are used in the circuit that contains the ignition switch operated by the driver. Much thicker wires are used in the circuit containing the battery and the high-current starter motor. Turning the ignition key causes a current to flow through the relay coil. This creates a magnetic field which attracts the armature to the core. The armature pivots and connects the starter motor to the car battery.

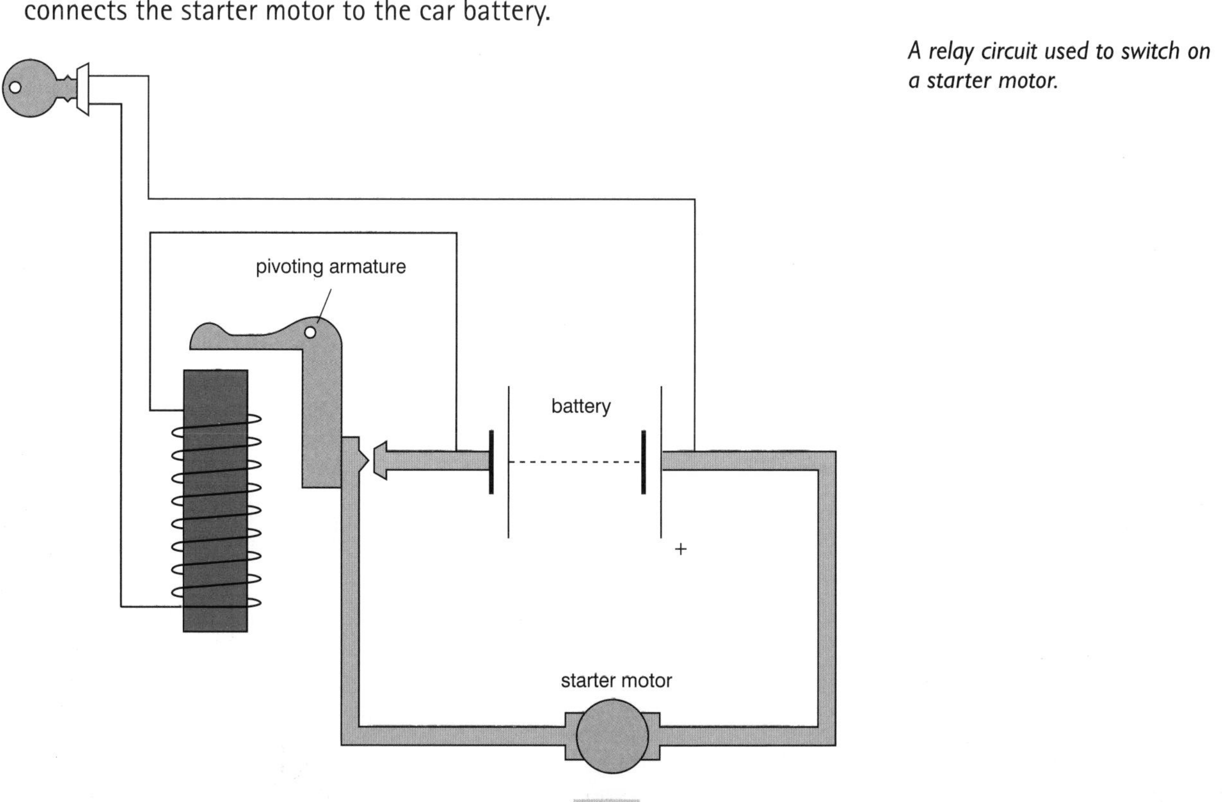

A relay circuit used to switch on a starter motor.

- Common examination questions show you a drawing of a d.c. electric motor and ask you to label the various parts and explain their functions.

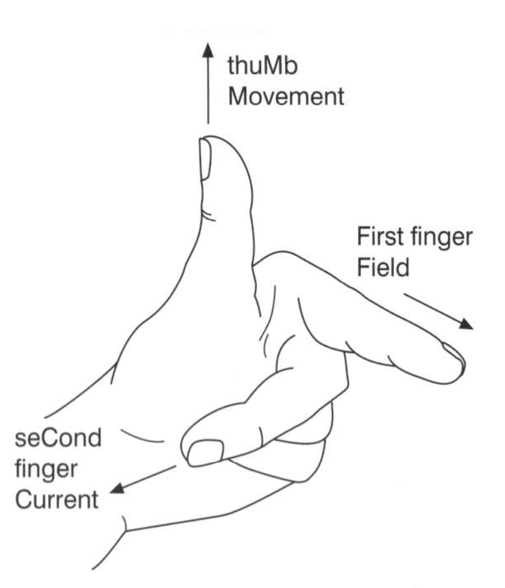

Fleming's left-hand rule predicts the direction of the force on a current carrying wire.

D.C. Motors

- An **electric motor** transfers electrical energy to kinetic energy. It is made from a coil of wire positioned between the poles of two permanent magnets. When a current flows through the coil of wire, it creates a magnetic field, which interacts with the magnetic field produced by the two permanent magnets. The two fields exert a force that pushes the wire at right angles to the permanent magnetic field.

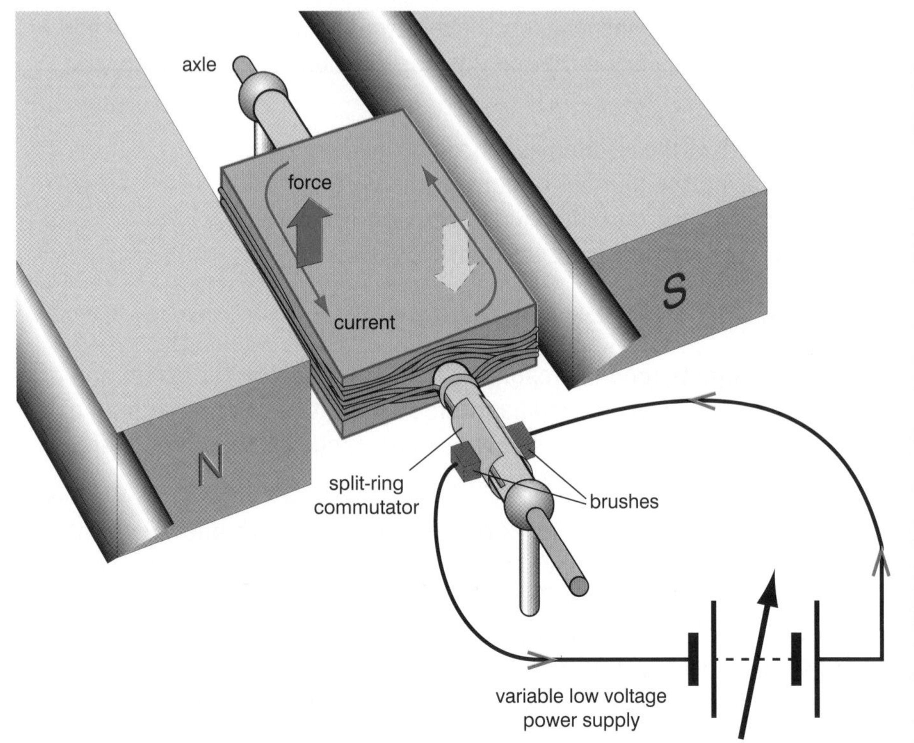

- A motor coil as set up in the diagram will be forced round as indicated by the arrows (1 and 2 below). The split-ring commutator ensures that the motor continues to spin. Without the commutator, the coil would rotate 90° and then stop. This would not make a very useful motor! The commutator reverses the direction of the current at just the right point (3) so that the forces on the coil flip around and continue the rotating motion (4).

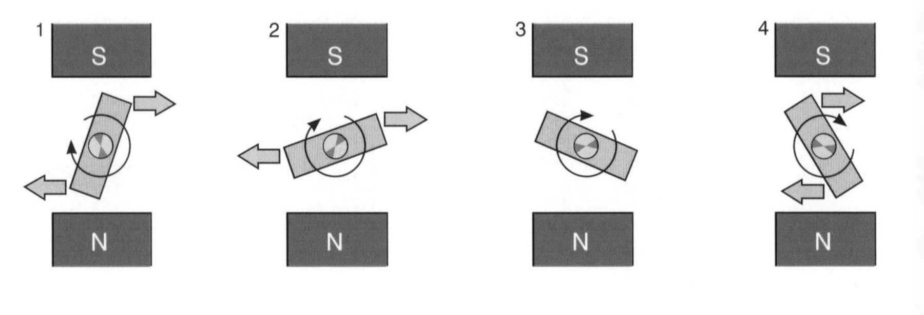

A* EXTRA

- Fleming's left-hand rule relates the movement of a coil to the direction of the permanent magnetic field and the direction of the current flowing in the coil.

QUESTION SPOTTER

- Questions often refer to 'soft' iron. Don't be confused – in this case 'soft' means that the magnetism is *temporary* in iron. Steel, for example, is a 'hard' magnetic material – it *stays* magnetised.

Check Yourself Questions

Q1 In a recycling plant, an electromagnet separates scrap metal from household rubbish.

a Can this method be used to separate aluminium drinks cans from other household rubbish? Explain your answer.

b How can the operator drop the scrap metal into the skip?

Q2 The diagram shows a simple electromagnet made by a student.

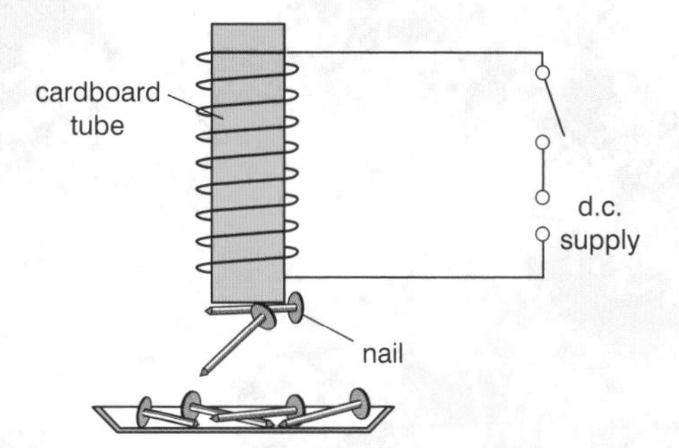

Suggest two ways in which the electromagnet can be made to pick up more nails.

Q3 The diagram shows an electric bell.

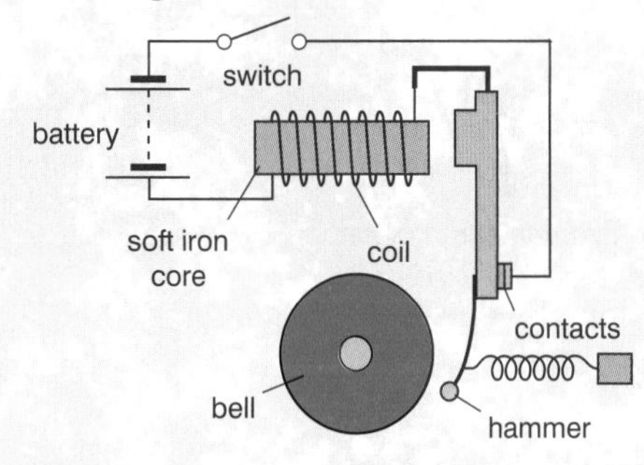

Explain how the bell works when the switch is closed.

Answers are on page 136.

Electromagnetic induction

Generating electricity

- Michael Faraday was the first person to generate electricity from a magnetic field using **electromagnetic induction.** The large generators in power stations generate the electricity we need using this process.
- Current is created in a wire when:
  - the wire is moved through a magnetic field ('cutting' the field lines)
  - the magnetic field is moved past the wire (again 'cutting' the field lines)
  - the magnetic field around the wire changes strength.
- Current created in this way is said to be **induced**.
- The faster these changes, the larger the current.

Dynamos and generators

- A **dynamo** is a simple current generator. A dynamo looks very like an electric motor. Turning the permanent magnet near to the coil induces a current in the wires. The split-ring commutator ensures that the current generated flows in only one direction.

In a bicycle dynamo, the magnet rotates and the coil is fixed.

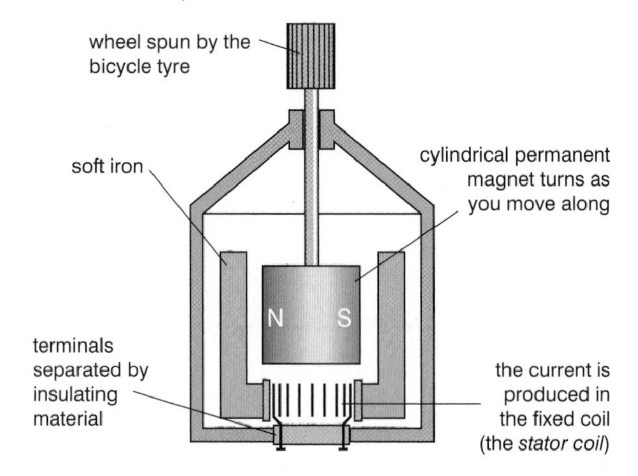

- Power station generators do not have a commutator, so they produce **alternating current**. Power stations use electromagnets rather than permanent magnets.

The generator rotates at 50 times per second, producing a.c. at 50 hertz.

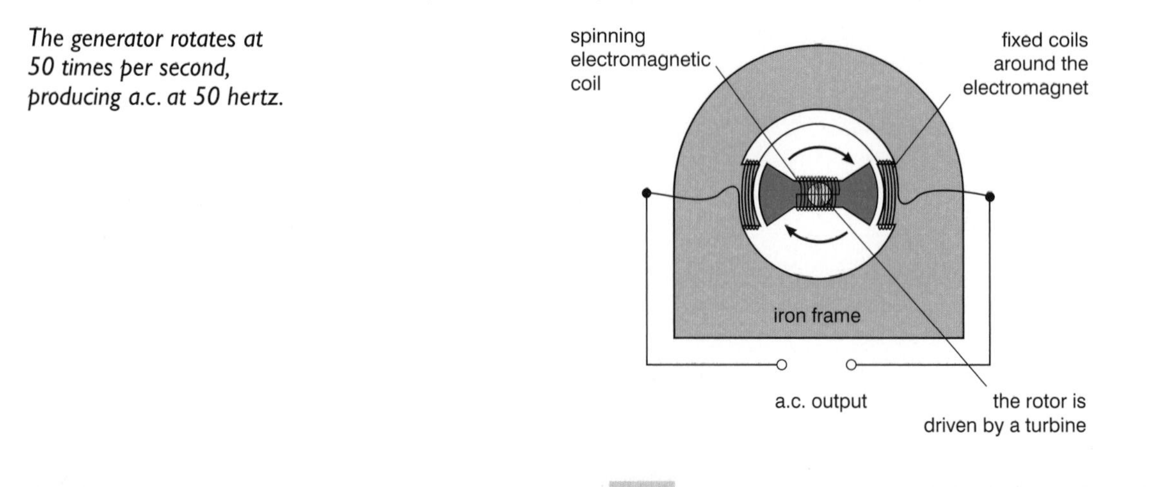

Transformers

- A **transformer** consists of two coils of insulated wire wound on a piece of iron. If an alternating voltage is applied to the first (primary) coil, the alternating current produces a changing magnetic field in the core. This changing magnetic field induces an alternating current in the second (the secondary) coil.
- If there are more turns on the secondary coil than on the primary coil, then the voltage in the secondary coil will be greater than the voltage in the primary coil. The exact relationship between turns and voltage is:

$$\frac{\text{primary coil voltage } (V_p)}{\text{secondary coil voltage } (V_s)} = \frac{\text{number of primary turns } (N_p)}{\text{number of secondary turns } (N_s)}$$

- When the secondary coil has more turns than the primary coil, the voltage increases in the same proportion. This is a **step-up transformer.**
- A transformer with fewer turns on the secondary coil than on the primary coil is a **step-down transformer**, which produces a smaller voltage in the secondary coil.

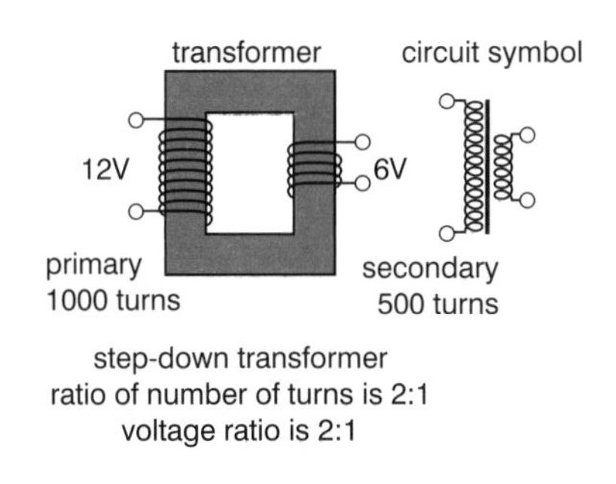

step-down transformer
ratio of number of turns is 2:1
voltage ratio is 2:1

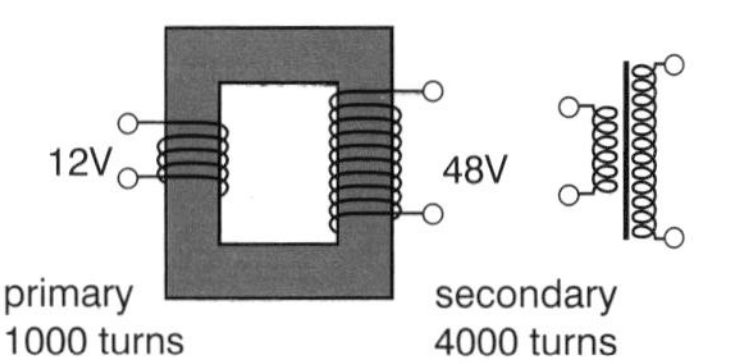

step-up transformer
ratio of number of turns is 1:4
voltage ratio is 1:4

Transformers are widely used to change voltages. They are frequently used in the home to step down the mains voltage of 230 V to 6 V or 12 V.

A* EXTRA

- In dynamos and transformers, changing the magnetic field around a coil of wire can create or induce an electric current in the coil. The direction of this induced current can be predicted using Fleming's right-hand rule.

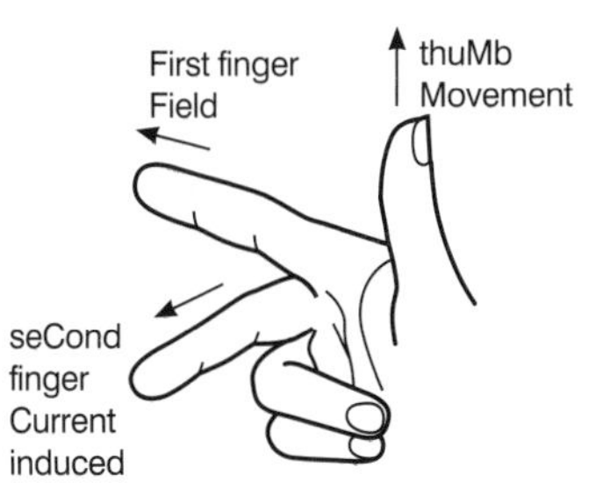

Fleming's right-hand rule predicts the direction of the current induced in a moving wire.

WORKED EXAMPLE

Calculate the output voltage from a transformer when the input voltage is 230 V and the number of turns on the primary coil is 2000 and the number of turns on the secondary coil is 100.

Write down the formula:	$\frac{V_p}{V_s} = \frac{N_p}{N_s}$
Substitute the values known:	$\frac{230}{V_s} = \frac{2000}{100} = 20$
Rewrite this so that V_s is the subject:	$V_s = \frac{230}{20}$
Work out the answer and write down the unit:	$V_s = 11.5\text{ V}$

- Calculations involving transformers are very common. Most often you will have to calculate output voltages from input voltages. Remember the equation because it will not be given in the examination.

Transmitting electricity

- Most power stations **burn fuel** to heat water into high-pressure steam, which is then used to drive a **turbine**. The turbine turns an a.c. generator, which produces the electricity.

The most common fuels used in power stations are still coal, oil and gas.

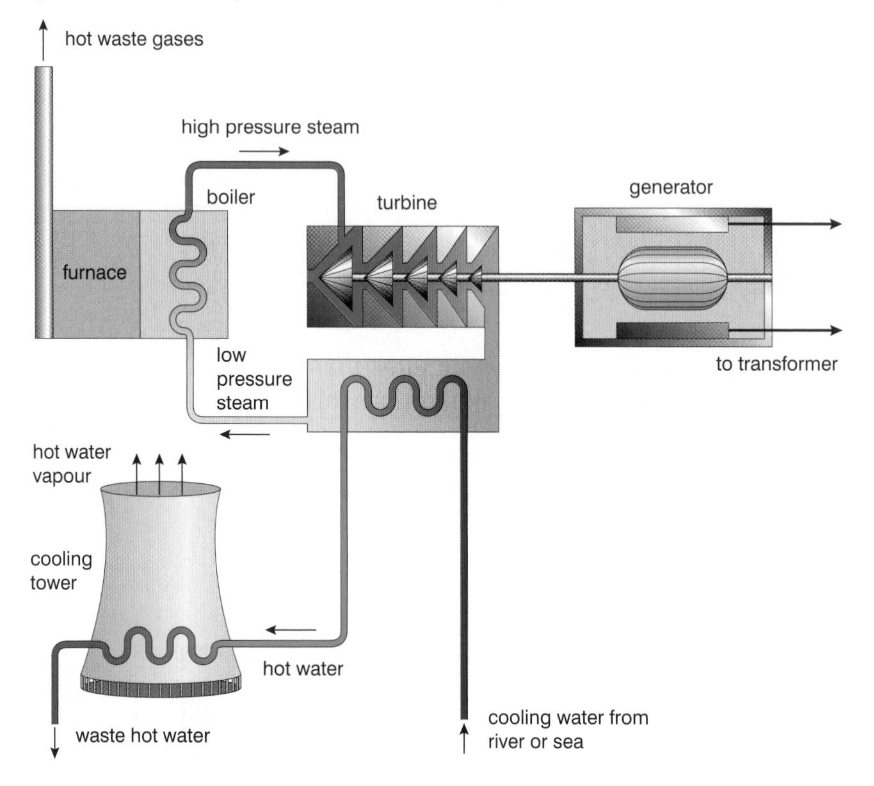

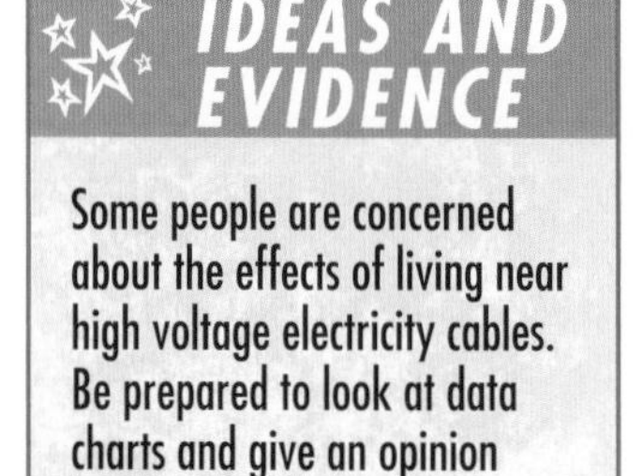

IDEAS AND EVIDENCE

Some people are concerned about the effects of living near high voltage electricity cables. Be prepared to look at data charts and give an opinion based on the information provided.

QUESTION SPOTTER

- Questions on this topic often involve an explanation with a number of steps. This makes it a good place to test your extended writing. Make sure you learn all the steps involved in, for example, the operation of transformers, generators, motors (page 16) and relays (page 15).

- The **National Grid** links all the power stations to all parts of the country. To minimise the power loss in transmitting electricity around the grid, the current has to be kept as low as possible. The higher the current, the more the transmission wires will be heated by the current and the more energy is wasted as heat.
- This is where transformers are useful. This is also the reason that mains electricity is generated as alternating current. When a transformer steps up a voltage, it also steps down the current and vice versa. Power stations generate electricity with a voltage of 25 000 V. Before this is transmitted on the grid, it is converted by a step-up transformer to 400 000 V. This is then reduced by a series of step-down transformers to 230 V before it is supplied to homes.

Mains electricity is a.c. so that it can be easily stepped up and down. High-voltage/low-current transmission lines waste less energy than low-voltage/high-current lines.

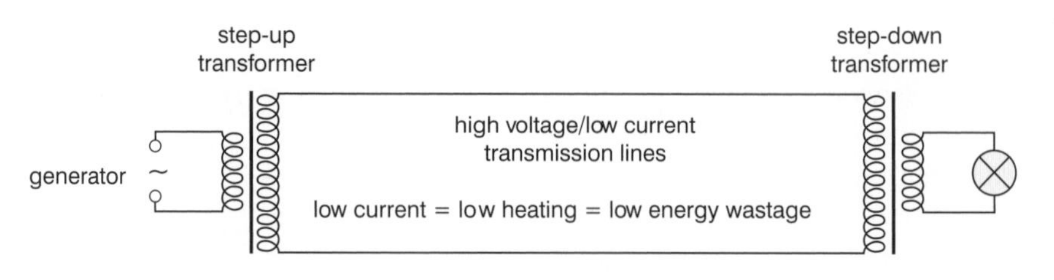

CHECK YOURSELF QUESTIONS

Q1 Two students are using the equipment shown in the diagram.

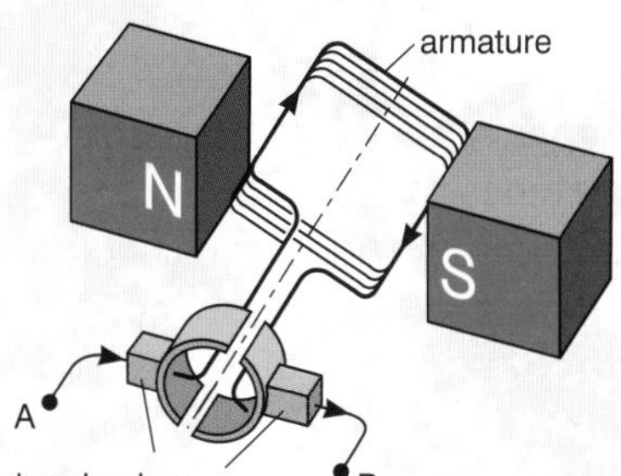

The students cannot decide whether it is an electric motor or a generator. Explain how you would know which it is.

Q2 The diagram shows a transformer.

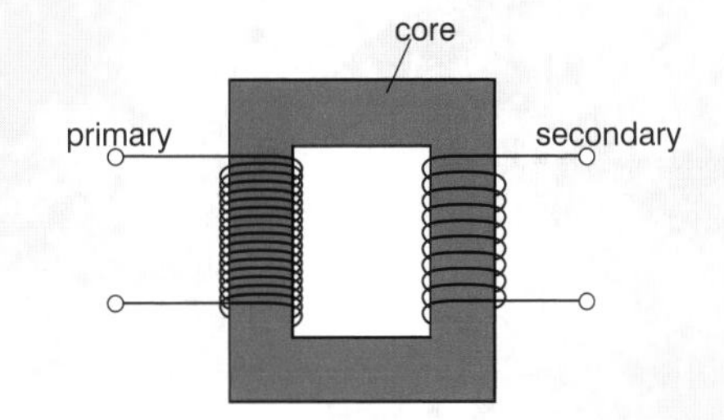

a What material is used for the transformer core?

b What happens in the core when the primary coil is switched on?

c What happens in the secondary coil when the primary coil is switched on?

d If the primary coil has 12 turns and the secondary coil has 7 turns, what will the primary voltage be if the secondary voltage is 14 V?

Q3 Explain why electricity is transmitted on the National Grid at very high voltages.

Answers are on page 137.

More electromagnetic devices

> **A* EXTRA**
>
> - Different diameter speakers are used for different frequencies of sound. This helps to *diffract* the different wavelengths more efficiently. (See page 76.)

How do loudspeakers work?

- Traditional loudspeakers make use of the **motor effect**, described in Revision Session 1.
- First, an amplifier circuit sets up a **varying electric current** in the coil. The variation in this current matches the variation in the sound wave that will be produced. When the current flows in the coil of wire, it creates a **magnetic field**, which interacts with the magnetic field produced by the permanent magnet. The two fields exert a **force** that pushes and pulls the coil at right angles to the permanent magnetic field. Because the coil is connected to the paper cone, the cone also moves and sets up a **vibration** in the air – a **sound wave**.

A moving coil loudspeaker.

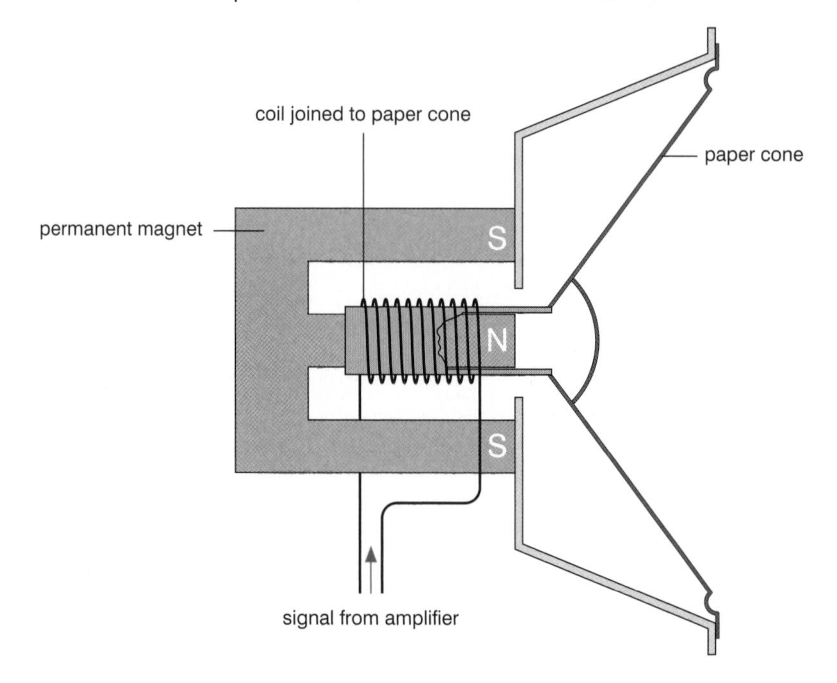

How do microphones work?

- Microphones transfer the energy from a sound wave – a pressure wave in the air – to a varying electrical signal. All microphones have a **diaphragm** that is vibrated by the incoming pressure wave. This vibration can be converted to an electrical signal in several ways.
- For GCSE Physics, we concentrate on the **moving-coil microphone**. This makes use of the generator effect described in Revision Session 2.
- In a moving-coil microphone, the incoming sound wave causes the diaphragm to vibrate. The diaphragm is connected to a **coil of wire**. This coil of wire vibrates in the **magnetic field** of a permanent magnet. The field lines of the magnet cut through the coil of wire as it moves, so a **varying electric current** is induced in the coil. This signal can then be amplified or stored in a number of ways.

> **A* EXTRA**
>
> - It is important to refer to coils of wire cutting across field lines or field lines cutting across a coil (generator effect). Anything that increases the rate of cutting (faster movement, more coils, stronger field) will increase the effect.

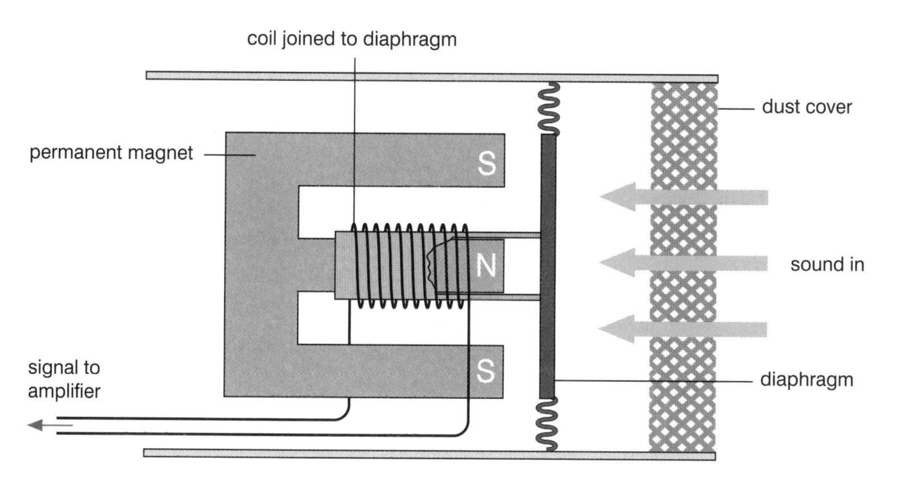

A moving coil microphone.

- Other types of microphone include carbon microphones, ribbon microphones, crystal microphones and condenser microphones.

How do tape recorders work?

- The tape heads in a cassette recorder work in one of two ways, depending on whether the recorder is being used in playback mode or record mode. During **recording**, the tape head uses the **motor effect** and during **playback**, it uses the **generator effect**.
- The head of a tape recorder consists of a very small, circular **electromagnet** with a small gap in it. During **recording**, the varying electric current in the wire generates a varying magnetic field around the head. While this happens, the tape is pulled past the tape head. The tape has a **chemical coating** (usually ferric oxide), which becomes **magnetised** by the electromagnet's field. The chemical 'lines up' in the varying field as it moves past.

QUESTION SPOTTER

- Be careful with the words you use. You would lose marks by saying 'the music is stored on tape'; it is the *information* that is stored as a pattern in the magnetic material.

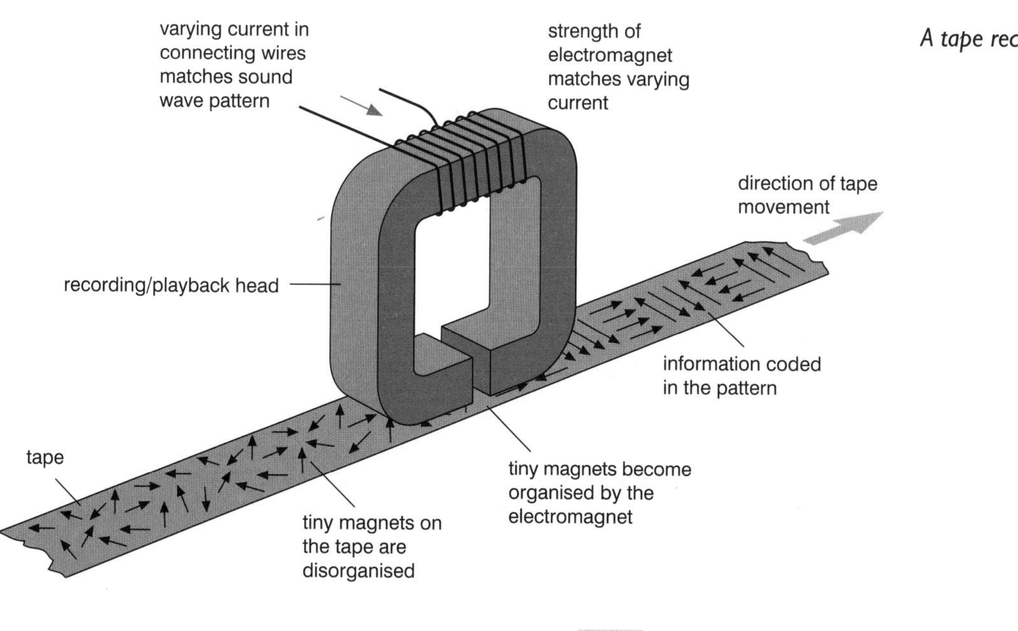

A tape recorder during recording.

- After the tape has moved on, the chemical stays in the 'lined up' position and so the patterns created by the varying magnetic field become stored on the tape. Because the varying magnetic field is caused by a current that follows the pattern of the music or sound being recorded, then the information stored on the tape also exactly follows the same pattern.
- During **playback**, the tape with its pattern of magnetic chemicals is pulled past the tape head. As the different patterns pass the coil, small varying currents are **induced** in the coil. These patterns exactly match the patterns of information stored on the tape. An **amplifier** increases these small currents to drive a loudspeaker and so the music or sound is recreated.

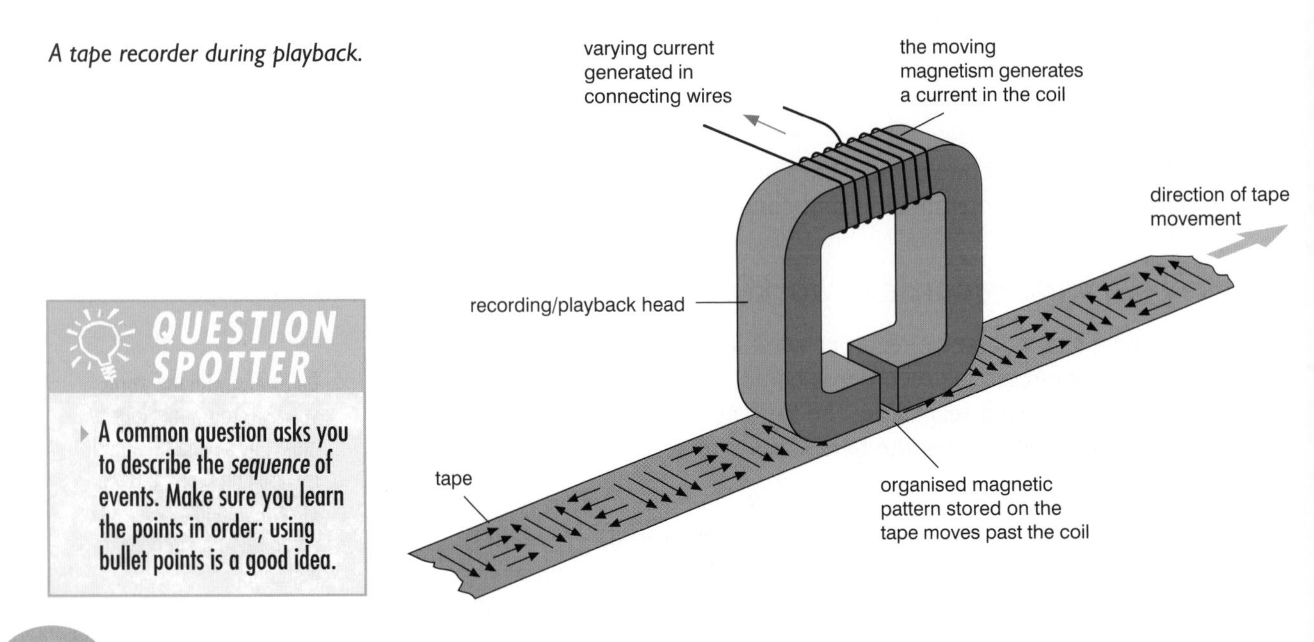

A tape recorder during playback.

QUESTION SPOTTER

- A common question asks you to describe the *sequence* of events. Make sure you learn the points in order; using bullet points is a good idea.

CHECK YOURSELF QUESTIONS

Q1 David is listening to his radio. He turns up the volume control. Describe the changes that occur which lead to him hearing a louder sound.

Q2 Helen has copied an audio tape several times and finds that the quality of the recording becomes worse each time. Explain why this happens. Use ideas about unwanted electrical signals (called 'noise') and think about how the pattern on the tape might be damaged.

Q3 Explain why it is generally not a good idea to store cassette tapes near to strong magnets (such as those in loudspeakers).

Answers are on page 137.

UNIT 3: ELECTRONICS AND CONTROL

REVISION SESSION 1 Logic gates

What are logic gates?

- A logic gate is an electronic circuit that has one or more **input** signals and one **output** signal. These signals are voltages that can be HIGH (about 5 V) or LOW (about 0 V). Logic gates are **digital** circuits as they can only have certain values of input and output – high or low. The output signal depends on the combination of signals at the inputs.

What are truth tables?

- Truth tables summarise the way in which a logic gate operates. Truth tables usually use 1 for a HIGH signal and 0 for a LOW signal.
- **AND** gate
  The output is high only when input A **AND** input B are high.

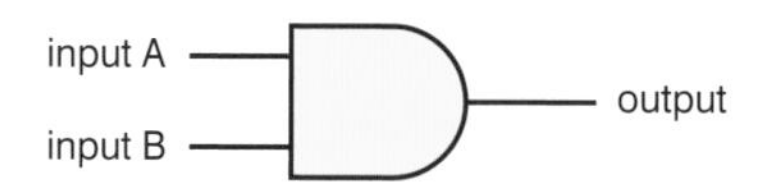

Inputs A	B	Output
0	0	0
0	1	0
1	0	0
1	1	1

- **OR** gate
  The output is high when input A **OR** input B is high, OR both.

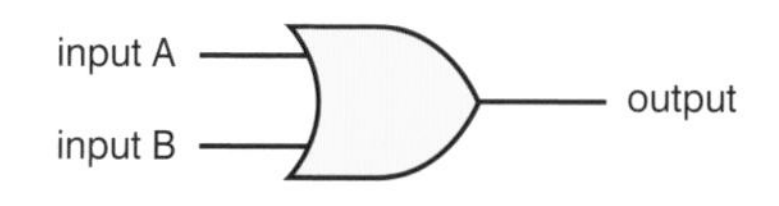

Inputs A	B	Output
0	0	0
0	1	1
1	0	1
1	1	1

- **NOT** gate
  This gate is also called an **inverter**. It has only one input.
  The output is high when the input is **NOT** high.

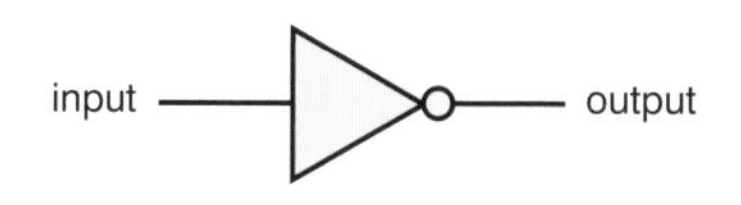

Input	Output
0	1
1	0

WORKED EXAMPLES

1 For safety, a car engine will not start unless the door is closed and the seat belt is fastened. Which type of logic gate is needed?

For the engine to start, both input conditions must be met. This circuit will need an AND gate.

2 A doorbell has switches at the front door and the back door of a house. Which type of logic gate is needed?

The bell needs to ring if either switch is pressed (or both). This circuit will need an OR gate.

More logic gates

- A **NOR** gate combines an OR gate and a NOT gate. The output is high when *neither* A NOR B are high.

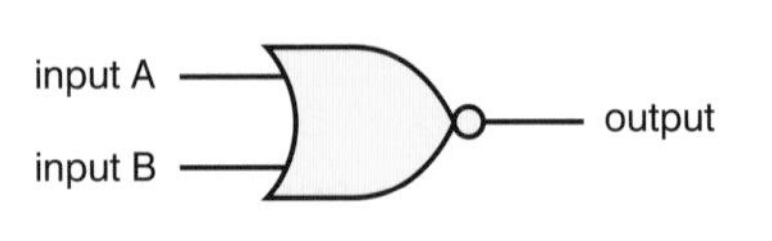

Inputs A	B	Output
0	0	1
0	1	0
1	0	0
1	1	0

A* EXTRA

- Logic gates have a **threshold** value – usually between 2 V and 3 V. Input voltages below the threshold value are treated as low (0 V) and voltages above the threshold are treated as high (5 V).

- A **NAND** gate combines an AND gate and a NOT gate. The output is high when *both* inputs are *not* high.

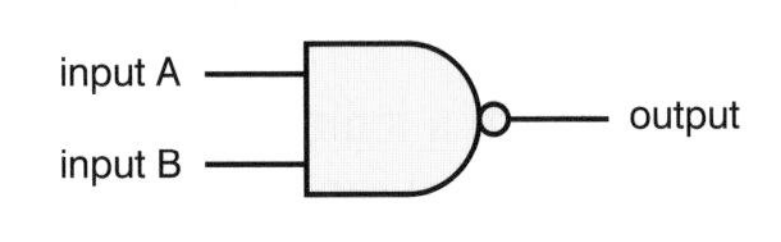

Inputs A	B	Output
0	0	1
0	1	1
1	0	1
1	1	0

Logic gate circuits

- Logic gates can be combined – the output signal from one gate can be used as the input signal to another.

WORKED EXAMPLE

Draw the truth table for this combination.

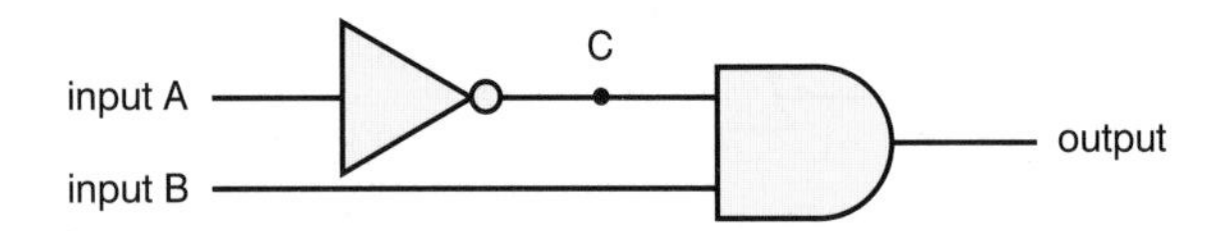

The key is to mark C on the diagram and work out this value first.

The value at C is NOT input A.

Inputs A	B	C	Output
0	0	1	
0	1	1	
1	0	0	
1	1	0	

The final output is input B AND the value at C.

Inputs A	B	C	Output
0	0	1	0
0	1	1	1
1	0	0	0
1	1	0	0

- For more complicated circuits, take it step by step and add 'way points' at the output of each logic gate.

Bistable circuits

- These circuits are made by cross-linking NOR gates or NAND gates.

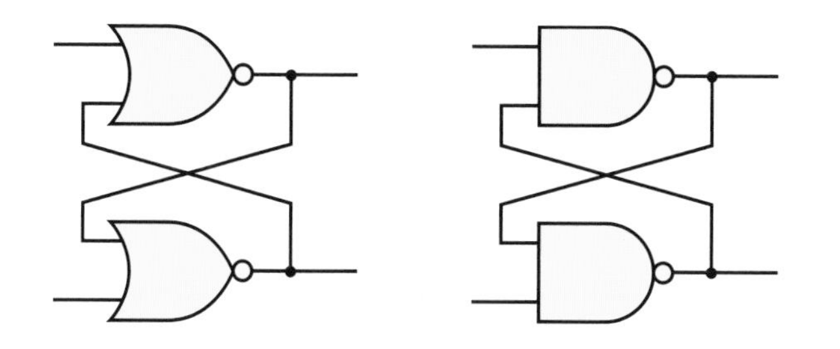

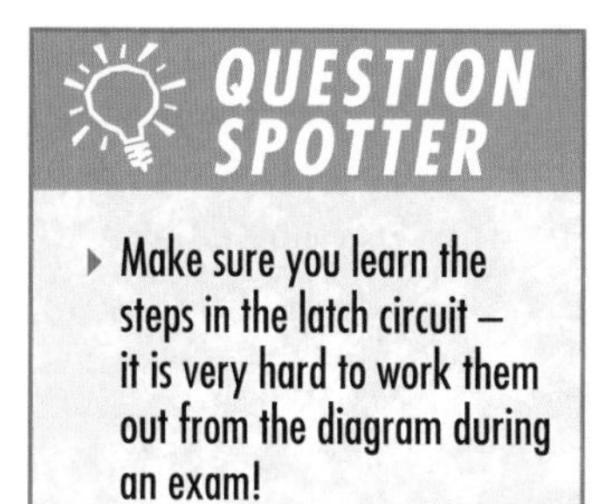

- The output depends on the **sequence** of changes at the inputs – the circuits act as a simple 'memory'. If only one output is used, the circuit is called a **latch**.
- In a NOR gate latch:

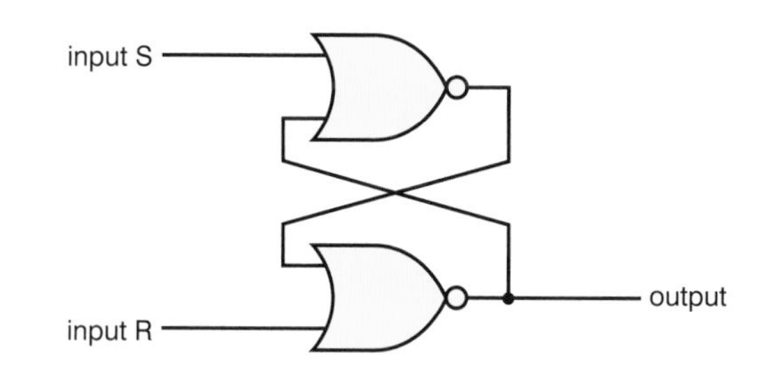

  - Initially, both inputs are low and the output is low.
  - Input S (the **set** input) goes high, so the output goes high.
  - Input S returns to low, but the output stays high – it is **latched on.**
  - Input R (the **reset** input) goes high, so the output returns to low.
  - Input R returns to low and the output stays low – it is **latched off**.
- This is useful in circuits such as burglar alarms where the input sensor (discussed in Revision Session 2) may only send a signal for a short time. The latch circuit keeps the alarm on until the reset is used.
- In a NAND gate latch, the sequence is the same except that:
  - Both inputs and the output are *high* at the start.
  - Moving the set input briefly to *low* changes the output to low.
  - Moving the reset input briefly to *low* returns the output to high.

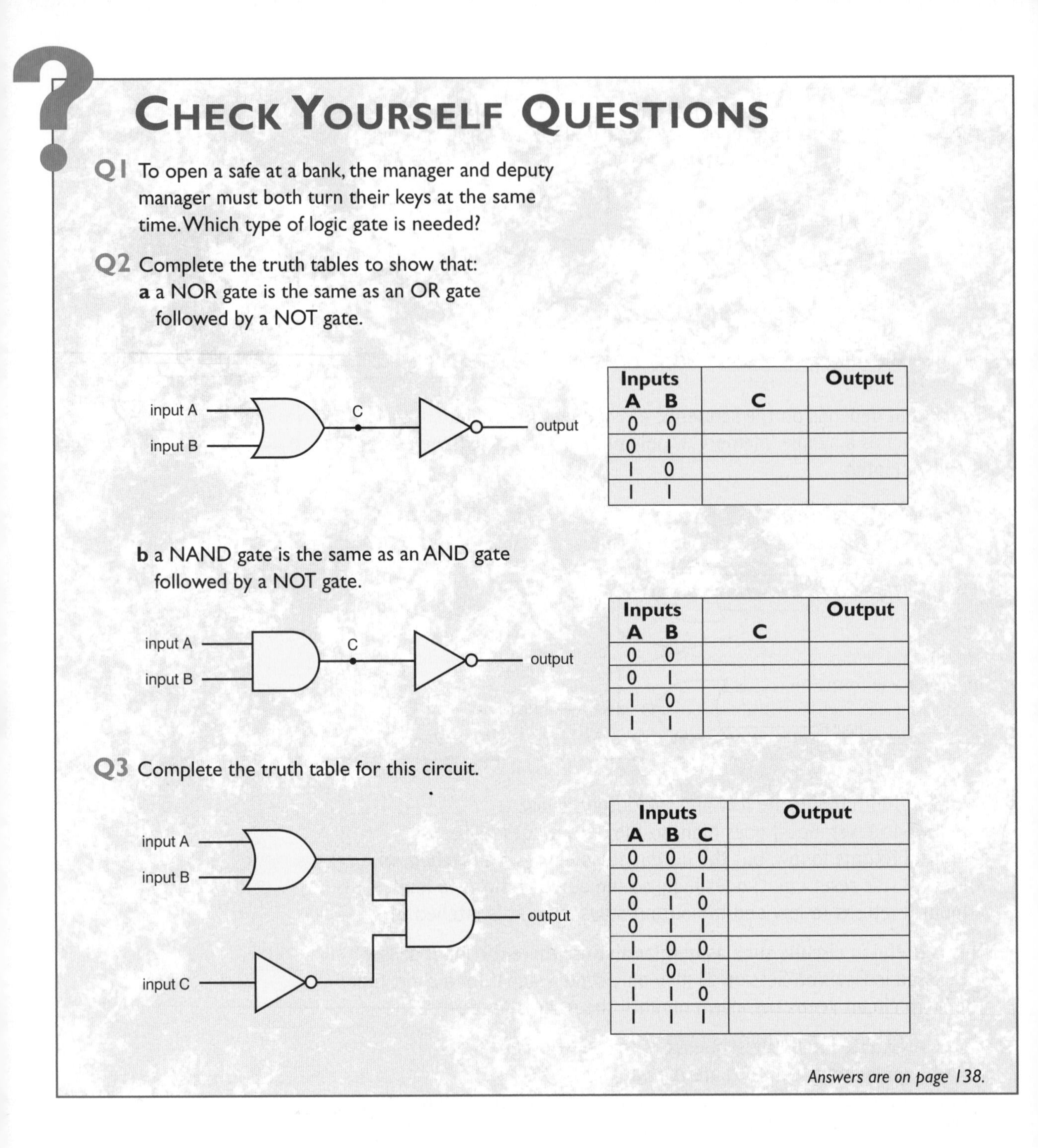

CHECK YOURSELF QUESTIONS

Q1 To open a safe at a bank, the manager and deputy manager must both turn their keys at the same time. Which type of logic gate is needed?

Q2 Complete the truth tables to show that:

a a NOR gate is the same as an OR gate followed by a NOT gate.

Inputs A	Inputs B	C	Output
0	0		
0	1		
1	0		
1	1		

b a NAND gate is the same as an AND gate followed by a NOT gate.

Inputs A	Inputs B	C	Output
0	0		
0	1		
1	0		
1	1		

Q3 Complete the truth table for this circuit.

Inputs A	Inputs B	Inputs C	Output
0	0	0	
0	0	1	
0	1	0	
0	1	1	
1	0	0	
1	0	1	
1	1	0	
1	1	1	

Answers are on page 138.

Input sensors and output devices

How do we provide signals for logic gates?

- The simplest way to provide an input signal for a logic gate is to use a switch and a resistor.

Switch	Input voltage to logic gate
open	LOW
closed	HIGH

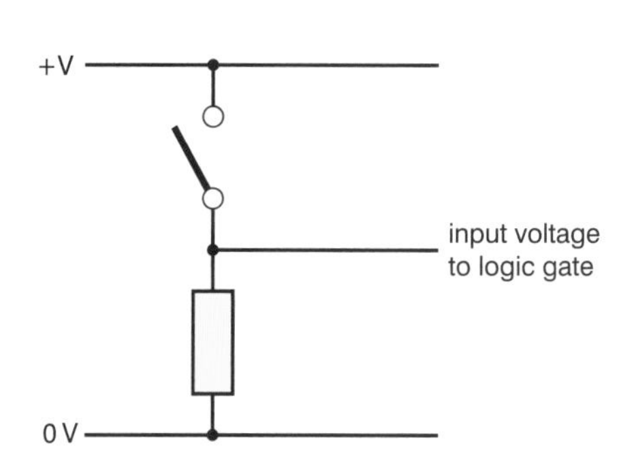

Sensors using potential dividers

- A sensor is a circuit in which a voltage changes as the environment changes. Most sensor circuits use a **potential divider**. A potential divider uses two resistors to split the supply voltage into two parts.
- The supply voltage is split in proportion to the values of the resistors. For example, if one resistor has twice the value of the other, its share of the voltage will be twice as high. By varying the resistances, we can vary the input voltage to the logic gate.

Calculating voltages in a potential divider

The current is the same in both resistors, so:

$$\frac{V_1}{V_2} = \frac{R_1}{R_2}$$

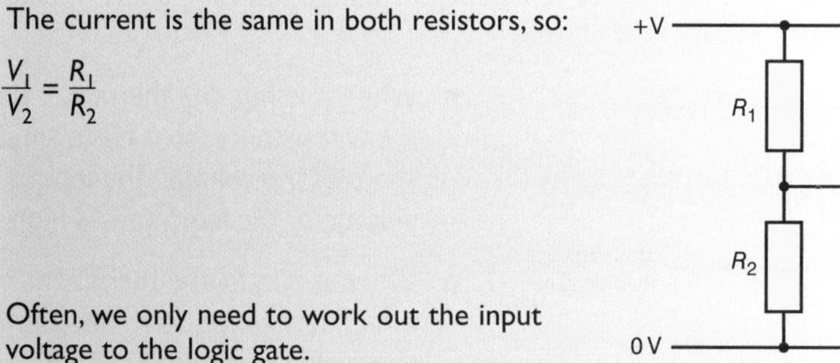

Often, we only need to work out the input voltage to the logic gate.

In this case, we can use the equation in this form:

$$V_{out} = \left(\frac{R_2}{R_1 + R_2}\right) V$$

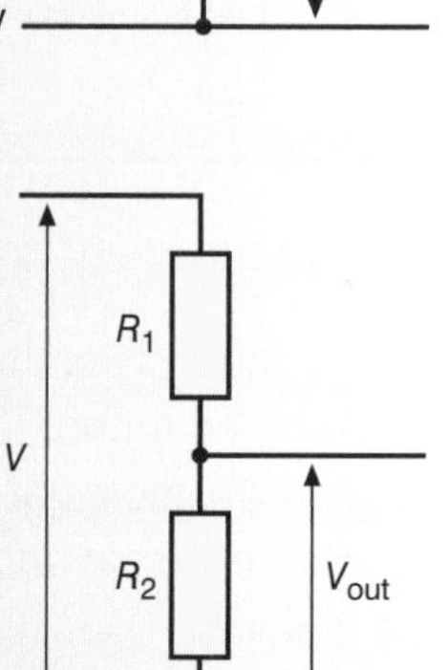

A* EXTRA

- Calculating the values of the voltage in a potential divider is tricky but becomes much easier with practice. See question 2 in the Check Yourself Questions.

Using a light-dependent resistor (LDR)

- The resistance of an **LDR** is high in the dark and low in the light.

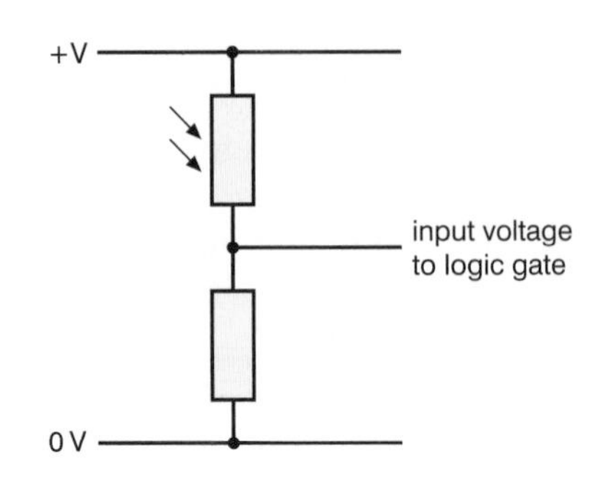

- When it is *light*, the LDR has a *low* resistance, so it has a small share of the voltage. The input voltage to the logic gate is high.
- When it is *dark*, the LDR has a *high* resistance, so it has a large share of the voltage. The input voltage to the logic gate is low.

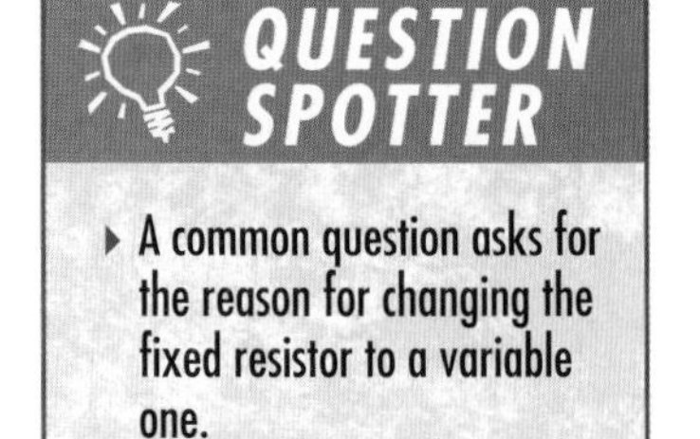

QUESTION SPOTTER

- A common question asks for the reason for changing the fixed resistor to a variable one.

- This time the resistors have been switched round. The input voltage to the logic gate will be high when it dark and low when it is light.
- If the resistor R is changed to a variable resistor, then we can vary the light level at which the input voltage to the logic gate goes from low to high.

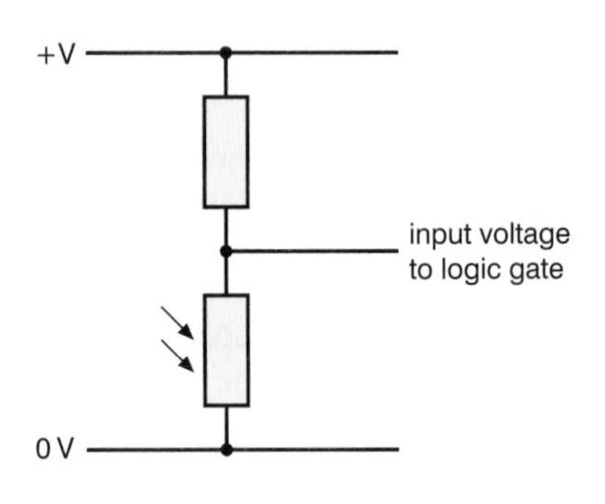

Using a thermistor

- The resistance of a **thermistor** is high at low temperatures and low at high temperatures.

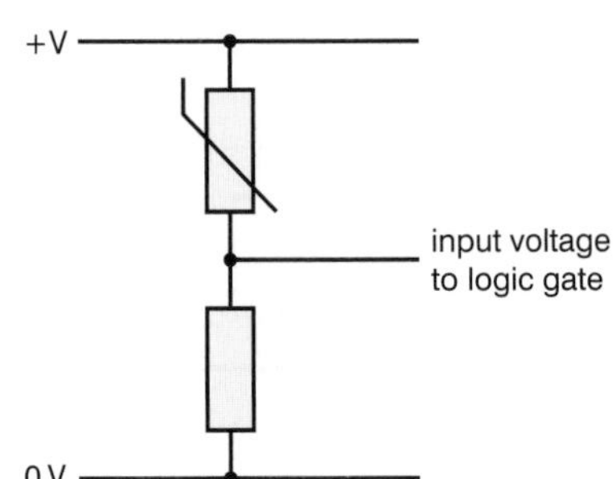
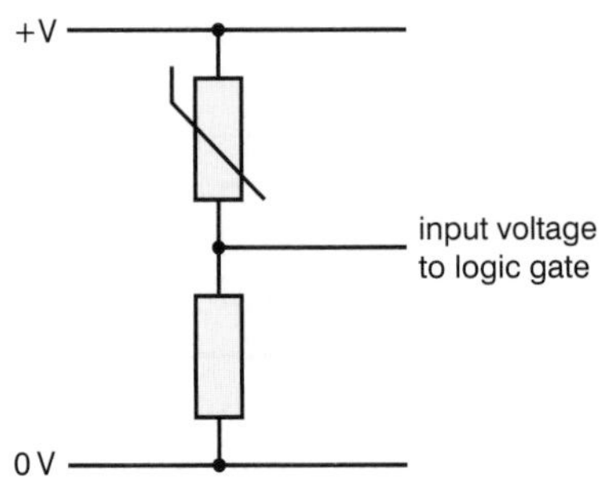

- When it is *hot*, the thermistor has a *low* resistance, so it has a small share of the voltage. The input voltage to the logic gate is high.
- When it is *cold*, the thermistor has a *high* resistance, so it has a large share of the voltage. The input voltage to the logic gate is low.

- This time the resistors have been switched round. The input voltage to the logic gate will be high when it is cold and low when it is hot.

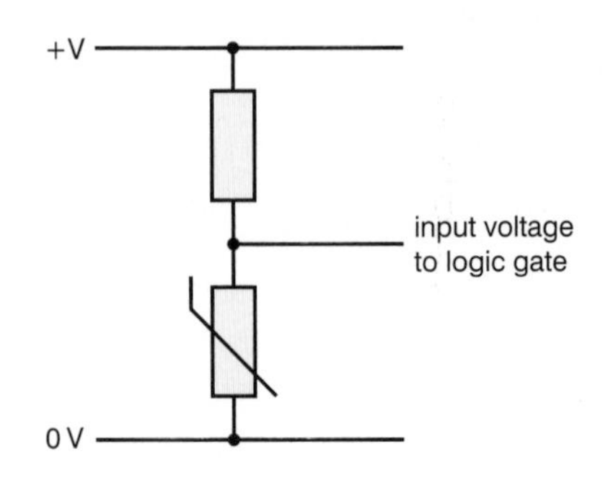

Other sensors

- A **moisture sensor** has two wire probes separated by a small gap.
- Moisture (water) conducts electricity better than the air, so the resistance changes when it is wet. This changes the input voltage to the logic gate.
- In a **tilt switch**, two contacts are only connected when the switch is turned in a particular direction. This changes the resistance, varying the input voltage to the logic gate.

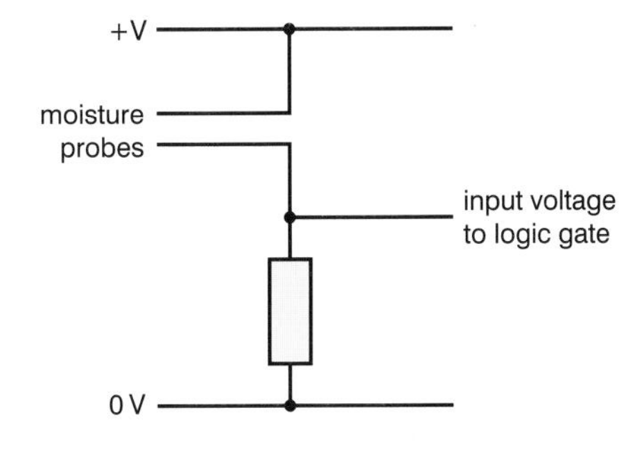

- Other sensors include:
  - **magnetic sensors** – a change in magnetic field causes a change in the input voltage to the logic gate
  - **pressure sensors** – the resistance of the sensor changes when the sensor is pressed in some way.

Output devices

- The output signal from a logic gate is large enough to drive devices such as **buzzers** and **LEDs (light-emitting diodes)**. When an LED is used, a resistor is used in series with it. This **protects** the LED against the current being too high.

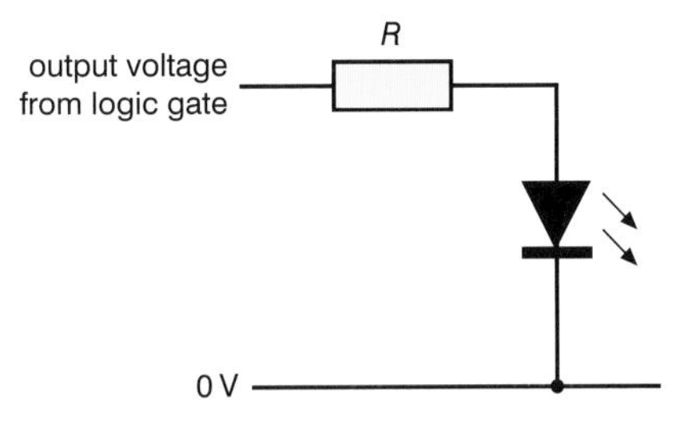

- The LED will light when the output from the logic gate is high (5 V).

WORKED EXAMPLE

When lit, an LED has a voltage of 1.5 V across it and a current of 10 mA.
Work out a suitable value for the protective resistor, *R*.

Voltage across resistor = output voltage from logic gate – voltage across LED
= 5 V – 1.5 V
= 3.5 V

Current through resistor = 10 mA = 0.01 A (because resistor and LED are in series)

Write Ohm's law in terms of *R*: $R = \frac{V}{I}$

Substitute the values for *V* and *I*: $R = \frac{3.5}{0.01}$

Work out the answer and write down the unit: $R = 350\ \Omega$

Using a relay

- The output signal from a logic gate is too small to operate devices such as motors, heaters and locks directly. To allow these devices to be connected, a **relay** is used.
- A relay is an electromagnetic switch.
- When the output from the logic gate is high, the relay coil is magnetised. This attracts the contacts and closes the switch, turning on the motor circuit. Relays are connected with a **protective diode** – this stops the logic gate being damaged when the relay coil demagnetises.

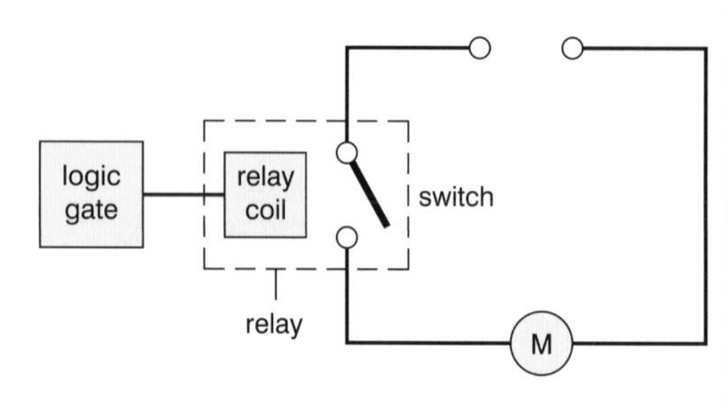

- A relay also **isolates** the logic gate, which could be damaged by large currents.

QUESTION SPOTTER

- Relays are an exam favourite. Make sure you know where they are used, how they operate and why they are used.

CHECK YOURSELF QUESTIONS

Q1 Here is a heat sensor circuit.

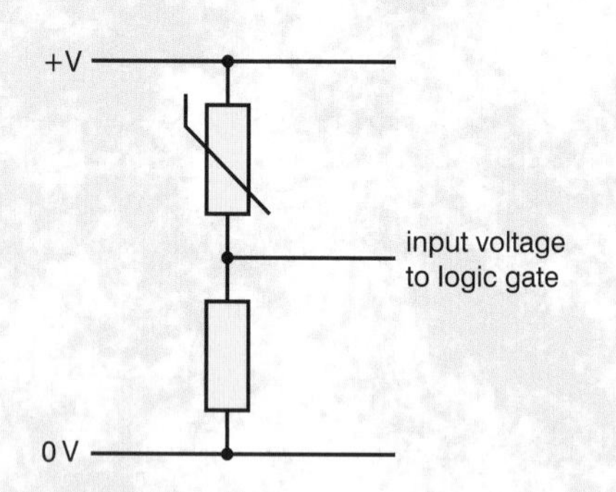

Describe and explain how the input voltage to the logic gate varies as the temperature changes from cold to hot.

Q2 Here is a light sensor circuit.

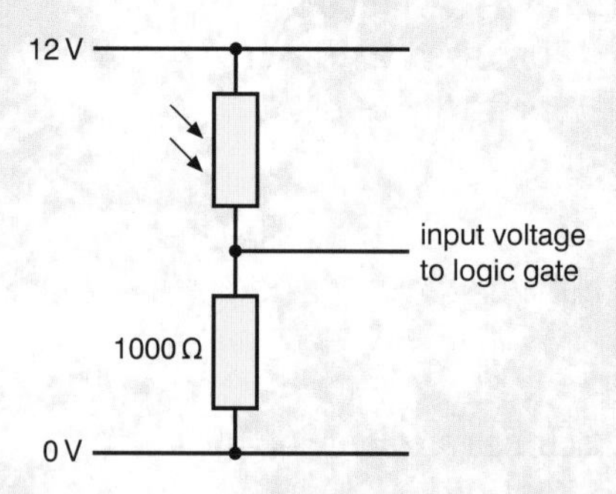

At a particular light level, the resistance of the LDR is 500 Ω. Calculate the input voltage to the logic gate.

Q3 State two reasons why a relay must be used when the output from the logic gate needs to control a heater.

Answers are on page 138.

REVISION SESSION 3

Electronic systems

How do the sections work together?

- Electronic systems are widely used. They have many benefits, but they do have disadvantages as well. The table shows some examples of this.

Electronic system	Advantages	Disadvantages
Mobile phone	• convenience • security in isolated places	• possible health hazards
Closed-circuit television (CCTV)	• improved security	• invasion of privacy
Internet	• access to worldwide information • ease of communication	• spread of unsuitable material • spread of computer viruses

- An electronic system consists of three parts:

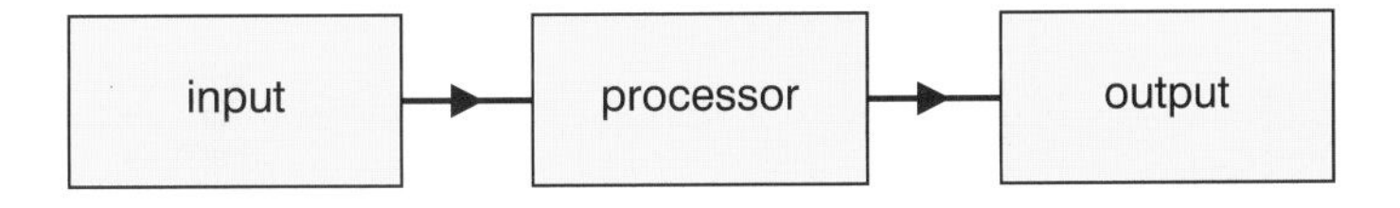

- The **input** section provides signals (voltages) for the processor. Often, these signals vary to reflect an environmental change.
- Examples of input devices are:
  - switches (such as push switches, magnetic switches, tilt switches)
  - heat sensors
  - light sensors
  - moisture sensors
  - pressure sensors.
- The **processor** section changes the input signal in a particular way.
- Examples of processor devices are:
  - electric circuits
  - transistors
  - logic gates.
- The **output** section includes the devices controlled by the processor section.
- Examples of output devices are:
  - LEDs
  - relays
  - buzzers
  - lamps
  - motors.

QUESTION SPOTTER

A common question gives you a block diagram of a system and then asks you to identify the input, processor or output sections.

QUESTION SPOTTER

- You might be given a partly complete diagram like this one and asked to draw in the missing bits – for example, to add the heat sensor potential divider.

WORKED EXAMPLE

Hannah designs an alarm for her fish tank. She wants an LED to light when the water gets too cold. She also wants a master switch to turn the system on.

Here is a block diagram of her system.

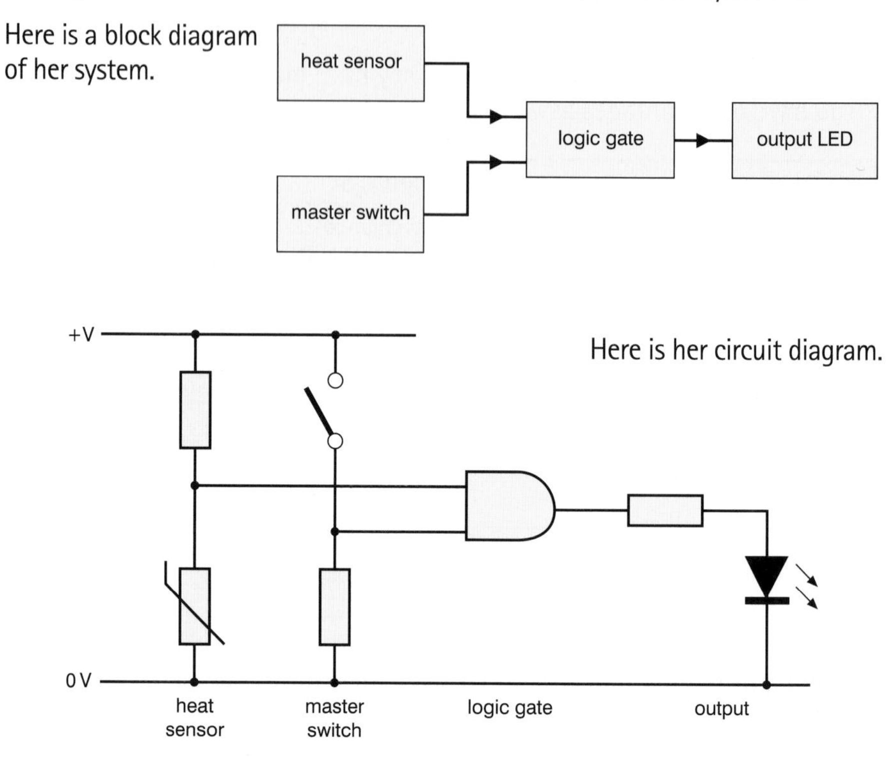

Here is her circuit diagram.

Describe the function of each part of her circuit.

Starting with the inputs of the system:

- The **heat sensor** is a potential divider. With the thermistor at the 'bottom', the signal to the logic gate will be high when the temperature is low – this is what she wants.
- The **master switch** provides a high input signal to the logic gate when she closes the switch – this is what she wants.

Looking at the processor in her system:

Hannah wants the LED to light when both inputs are high, so her **processor** is an AND gate.

Looking at the output of her system:

The LED (and its protective resistor) is the **output** for the system.

A* EXTRA

- Be prepared to suggest suitable values for the resistors. You may need to use potential divider formulas and/or Ohm's law: $V = IR$

QUESTION SPOTTER

- It is common for a question to set up a situation and ask you to complete the desired outcomes, e.g. a bell rings when a burglar stands on a mat in the dark. This is to help you to identify what the system should do. Only later in the question will it ask about particular parts of circuits.

Designing a system

- These are the stages in designing a system:
  - Decide what the system is supposed to do – what output do you want?
  - Break the system into input, processing and output tasks.
  - Put the system together as a block diagram.
  - Choose circuits for each of the blocks.
  - Connect the blocks together.

A* EXTRA

- You will have, at most, three inputs in any system you have to analyse or design.

Capacitors

- A capacitor is a device that can store electric charge (usually electrons).
- As a capacitor charges up, the voltage across it increases. When a capacitor is used with a resistor in a potential divider circuit, it acts as a **time delay**.

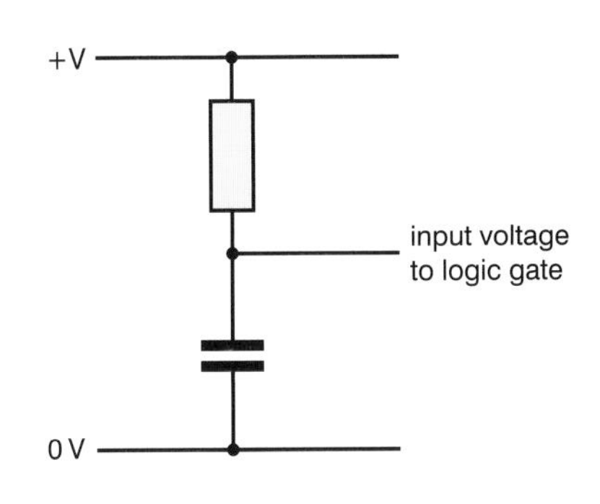

- The output voltage from the potential divider increases as the capacitor charges up. If this output voltage is used as the input voltage to a logic gate, the logic gate input goes from low to high after a certain time delay.
- The length of the time delay depends on the value of:
  - the **resistance** – a larger value gives a longer time delay
  - the **capacitance** – a larger value gives a longer time delay
  - the **supply voltage** – a larger value gives a shorter time delay.
- If the positions of the resistor and capacitor were reversed, then the input voltage to the logic gate would start high and become low after a delay.

A system using a transistor as a processor

- In this circuit:
  - The LDR and the variable resistor are a potential divider providing the input voltage for the NOT gate. When it is dark, the resistance of the LDR is high so it takes a larger share of the voltage – the input voltage to the NOT gate will be low. The exact light level at which low becomes high can be set by adjusting the variable resistor.

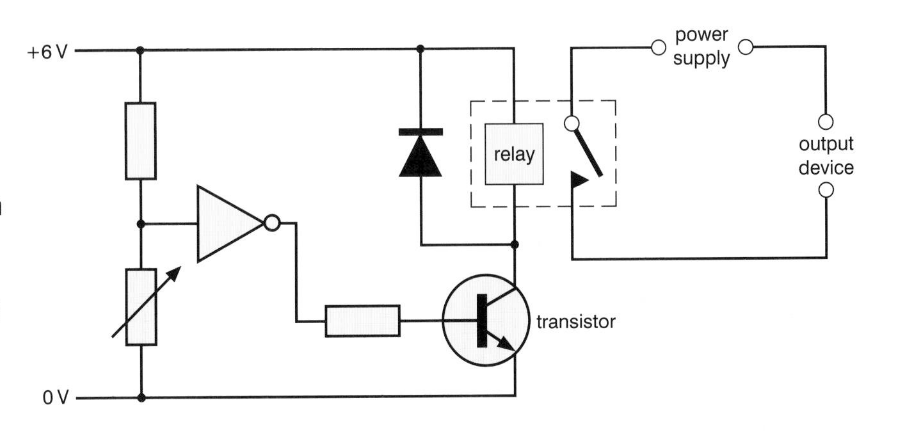

  - The NOT gate is the **processor**. When it is dark, the input to the NOT gate is low, so the output will be high.
  - The transistor acts as a switch. When the output from the NOT gate is high, the transistor will allow a current to flow through the relay coil. The diode protects the transistor from damage when the coil demagnetises.
  - The magnetised relay coil will close the contacts in the second circuit, switching on the output device.
- Overall, the output device is switched on when the light reaches a sufficient level of brightness. This could be useful in, for example, turning on a motor to automatically close some blinds when the light level is too high.

CHECK YOURSELF QUESTIONS

Q1 Pardeep has built an alarm for his bedroom.

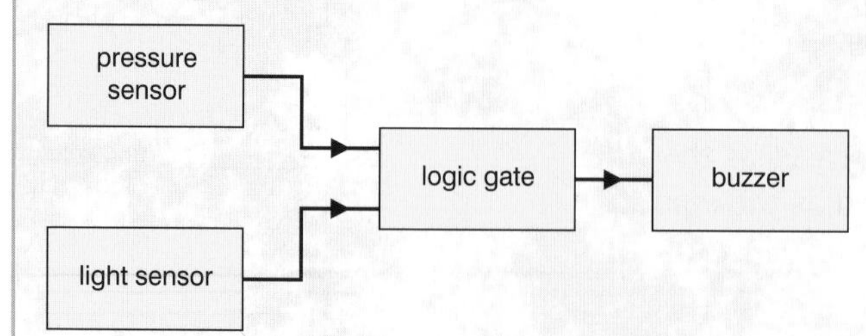

a Identify the input, processing and output devices.

b Pardeep wants the buzzer to sound if the pressure sensor sends a high signal and the master switch sends a high signal. Which logic gate should he use?

c Pardeep find that the buzzer goes off as soon as any 'burglar' steps off the pressure sensor. How could Pardeep change his system so that the alarm stays on until the master switch is pressed?

d Instead of sounding a buzzer, Pardeep decides that the system should activate a door lock. To do this he needs to use a relay to activate the lock circuit. Redraw the block diagram to include these features.

Q2 Annie is making a model set of traffic lights using LEDs. To make the sequence of colours work correctly she is using a logic circuit. Here is her diagram.

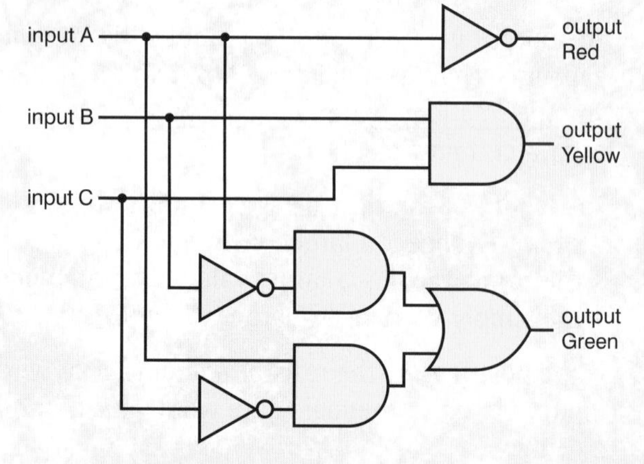

Copy and complete the truth table.

Inputs			Outputs		
A	**B**	**C**	**R**	**Y**	**G**
0	0	0			
0	0	1			
0	1	0			
0	1	1			
1	0	0			
1	0	1			
1	1	0			
1	1	1			

Answers are on page 138.

UNIT 4: FORCES AND MOTION

REVISION SESSION 1

The effects of forces

How do forces act on objects?

- It is very unusual for a single force to be acting on an object. Usually there will be two or more. The size and direction of these forces determine whether the object will move and the direction it will move in.
- Forces are measured in **newtons**. They take many forms and have many effects including pushing, pulling, bending, stretching, squeezing and tearing. Forces can:
  - **change the speed** of an object
  - **change the direction** of movement of an object
  - **change the shape** of an object.

What is friction?

- **Friction** is a very common force. It is the force that tries to stop movement between touching surfaces by opposing the force that causes the movement.
- In many situations friction can be a disadvantage, e.g. friction in the bearings of a bicycle wheel. In other situations, friction can be an advantage, e.g. between brake pads and a bicycle wheel.

The effects of weight

- **Weight** is another common force. It is also measured in **newtons**. The weight of an object depends on its **mass** and **gravity**. Any mass near the Earth has weight due to the Earth's gravitational pull.
- Weight is calculated using the equation:

weight = mass × gravitational field strength

- On the Moon, your mass will be the same as on Earth, but your weight will be less. This is because the gravity on the Moon is about one-sixth of that on the Earth, and so the force of attraction of an object to the Moon is about one-sixth of that on the Earth.

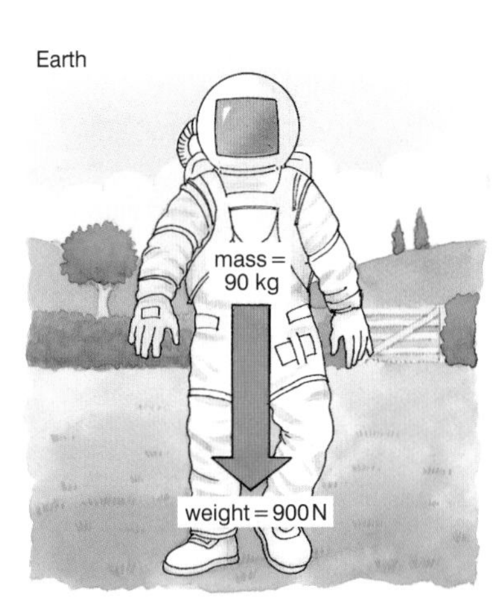

Who cares how much you weigh? It's your mass that people really care about.

Balanced forces

- Usually there are at least two forces acting on an object. If these two forces are **balanced** then the object will either be stationary or moving at a constant speed.
- A spacecraft in deep space will have no forces acting on it – no air resistance (no air!), no force of gravity – and because there is no need to produce a forward force from its rockets, it will travel at a constant speed.

Unbalanced forces

- For an object's speed or direction of movement to change, the forces acting on it must be **unbalanced**.

QUESTION SPOTTER

- You may be given diagrams of objects – with arrows showing the size and direction of the forces acting on them – and asked to say whether the object is moving or stationary.

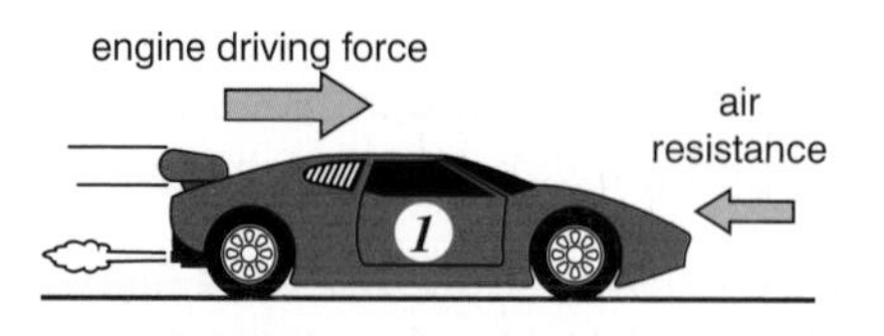

Acceleration. The force provided by the engine is greater than the force provided by air resistance and so the car increases its speed.

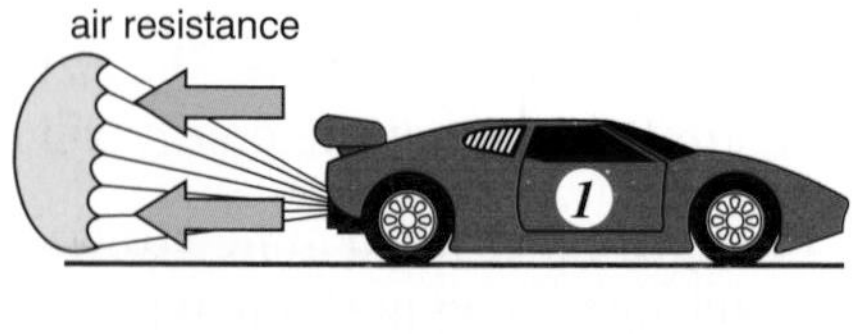

Deceleration. The engine is now providing no forward force. The 'drag' force provided by the parachute will slow the car down.

- As a gymnast first steps on to a trampoline, his weight is much greater than the opposing supporting force of the trampoline, so he moves downwards, stretching the trampoline. As the trampoline stretches, its supporting force increases until the supporting force is equal to the gymnast's weight. When the two forces are balanced, the trampoline stops stretching. If an elephant stood on the trampoline, it would break because it could never produce a supporting force equal to the elephant's weight.

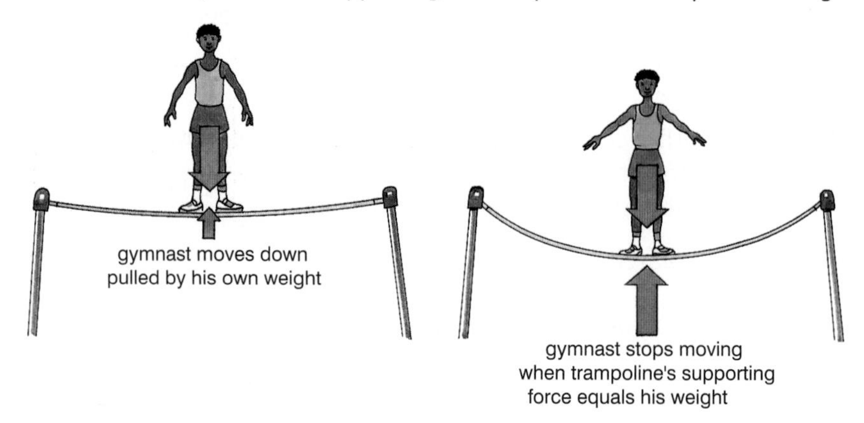

A trampoline stretches until it supports the weight on it.

A* EXTRA

- An object can be moving even when there is no resultant force acting on it. For example, there is no resultant force acting on a skydiver once the terminal speed has been reached.
- In deep space (well away from any planets) a spacecraft can travel at very high speeds without any force being applied from its rockets.

- As a skydiver jumps from a plane, the force of gravity will be much greater than the opposing force caused by air resistance. The skydiver's speed will increase rapidly – and the force caused by the air resistance increases as the skydiver's speed increases. Eventually it will exactly match the force of gravity, the forces will be balanced and the speed of the skydiver will remain constant. This speed is known as the **terminal speed**.
- If the skydiver spreads out so that more of his or her surface is in contact with the air, the resistive force will be greater and the terminal speed will be lower than if he or she adopted a more compact shape. A parachute has a very large surface, and produces a very large resistive force, so the terminal speed of a parachutist is quite low. This means that he or she can land relatively safely.

How are springs affected by stretching?

- When a spring stretches, the extension of the spring is proportional to the force stretching it, provided the elastic limit of the spring is not exceeded. This is **Hooke's law** and is shown by a straight line on a graph.

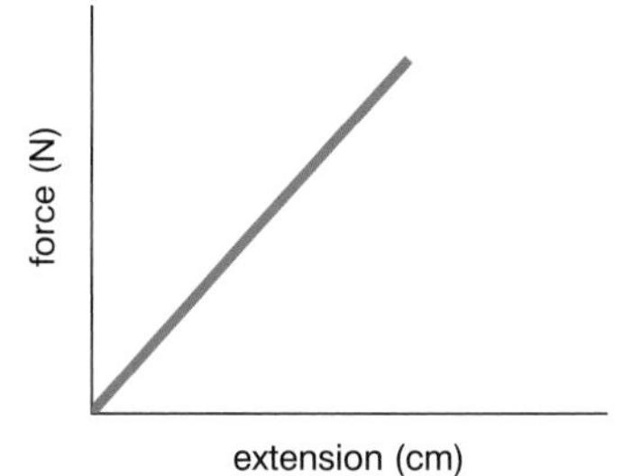

Elastic behaviour in a spring is shown by a straight line.

- The gradient of the line is a measure of the stiffness of the spring – the steeper the line, the stiffer the spring. This stiffness is the **spring constant**. As an equation, this is:

force = spring constant × extension

- When a spring is stretched, energy is transferred to it. The amount of energy stored in the spring is given by the area under the force–extension graph. If the graph is a straight line, the area under the graph is a triangle, so the energy is calculated as:

energy stored = $\frac{1}{2}$ × force × extension

- Hooke's Law shows **elastic** behaviour – when the force is removed, the spring returns to its original length.
- If the spring is stretched further, we obtain this graph:

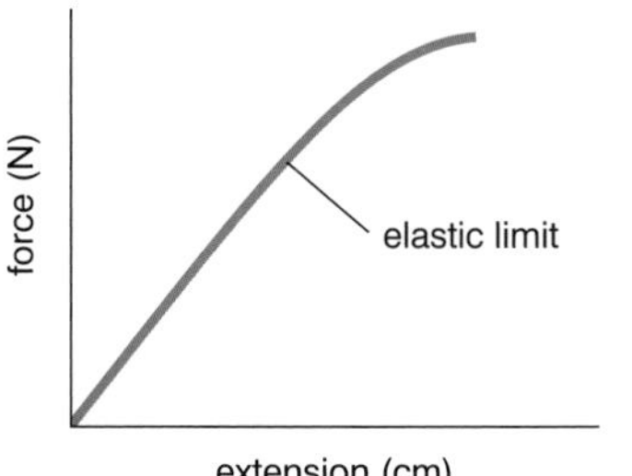

Plastic behaviour in a spring is shown by a smooth curve after the elastic limit has been reached.

- After the **elastic limit** has been reached, the spring starts to 'give' and is now **permanently** stretched – it will not go back to its original length when the force is removed. This is **plastic** behaviour.

A* EXTRA

- During elastic behaviour, the particles in the material are pulled apart a little. During plastic behaviour the particles slide past each other, and the structure of the material is changed permanently.

How are other materials affected by stretching?

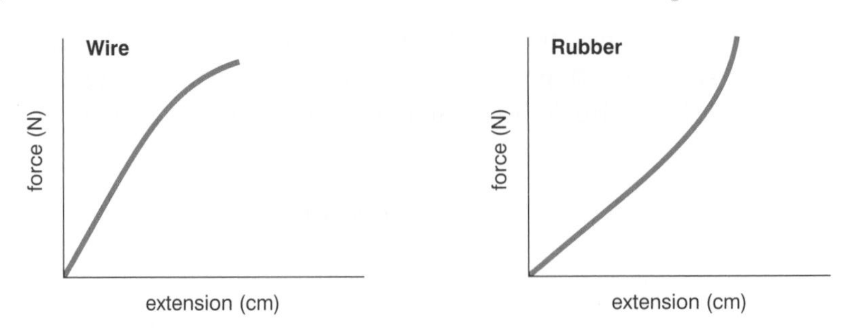

Force–extension graphs for a metal wire and for rubber.

- A metal wire has a large plastic section on the graph. As it stretches, the wire becomes thinner and thinner until it finally breaks.
- A piece of rubber stretches a quite a lot for small forces. The long polymer molecules are being 'straightened out'. Once this is done it becomes much stiffer and harder to extend further.

IDEAS AND EVIDENCE

As imaging and technological processes have improved, for example the very large magnifications given by electron microscopes, so many new materials have been developed. Working at very small scales has allowed computer chips to become smaller and much more powerful. Individual atoms are moved about as scientists begin to work at the scale of nanotechnology. These developments all stem from testing and measuring the properties of materials and building up a model of the internal structure of materials.

CHECK YOURSELF QUESTIONS

Q1 Look at the diagrams A, B and C. In each case describe the effect the forces would have on the object.

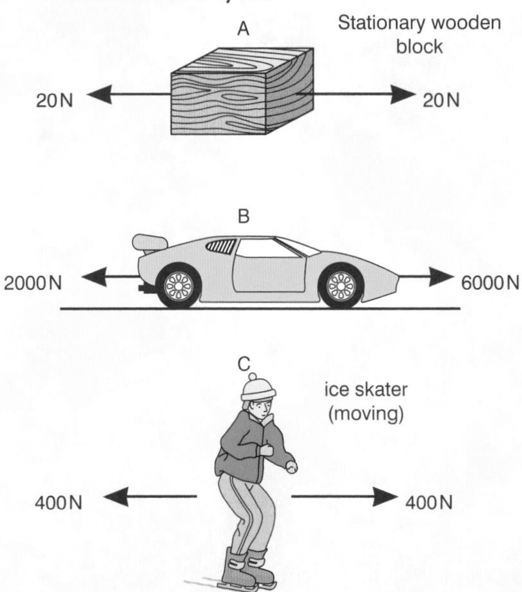

Q2 Skiers can travel at very high speeds. A friction force acts on the skis.

a Use the diagram to show the direction of the friction force acting on the skier.

b Explain why it is important for the friction forces to be as small as possible.

c Describe one way the skier could reduce the friction force.

Q3 The diagram shows the stages in the descent of a skydiver.

a Describe and explain the motion of the skydiver in each case.

b In stage 5 explain why the parachutist does not sink into the ground.

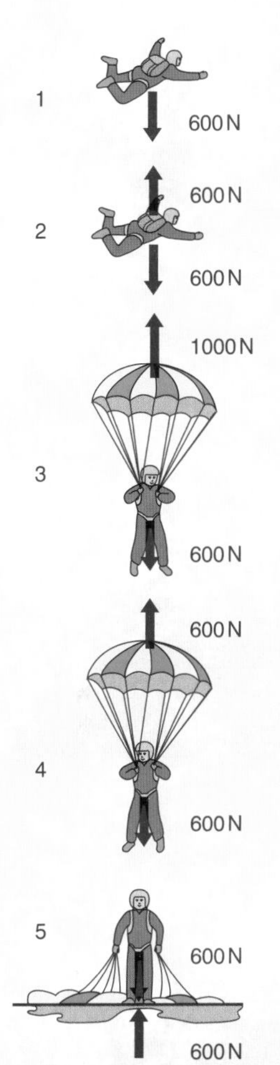

Q4 Terry performed an experiment stretching a spring. She loaded masses onto the spring and measured its extension. Here are her results.

Extension in cm	0	4	8	12	16	20	24
Load in N	0	2.0	4.0	6.0	7.5	8.3	8.6

a On graph paper, plot a graph of Load (y-axis) against Extension (x-axis). Draw a suitable line through your points.

b Mark on the graph the regions of elastic and plastic behaviour.

c Calculate the spring constant (stiffness) of the spring in the elastic region.

Answers are on page 139.

Velocity and acceleration

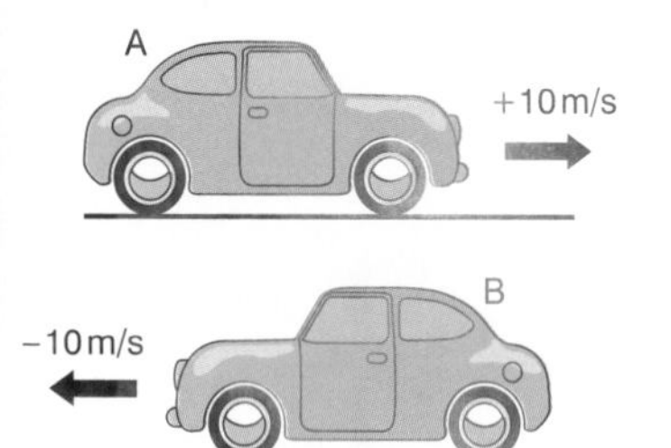

Both cars have the same speed. Car A has a velocity of +10 m/s, car B has a velocity of −10 m/s.

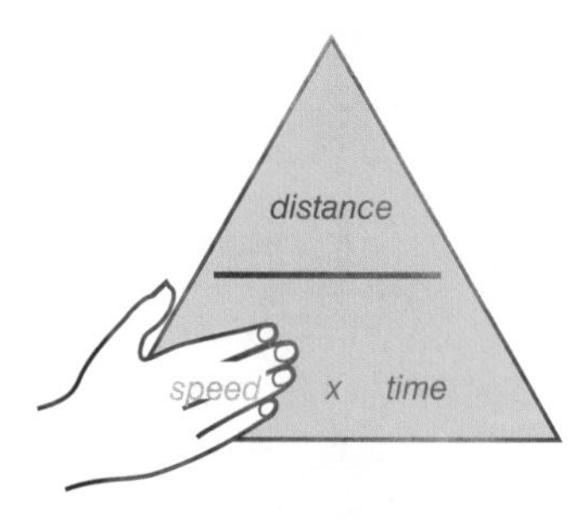

Cover speed to find that $speed = \frac{distance}{time}$

> **QUESTION SPOTTER**
>
> - Calculation questions involving $v = \frac{s}{t}$ are very common. You will need to remember the equation and be able to change the subject of the equation to match the question. (For example, write $s = v \times t$.) Don't forget to include the correct units.

> **A* EXTRA**
>
> - A negative acceleration shows that the object is slowing down.

Are speed and velocity the same?

- The **speed** of an object can be calculated using the following formula:

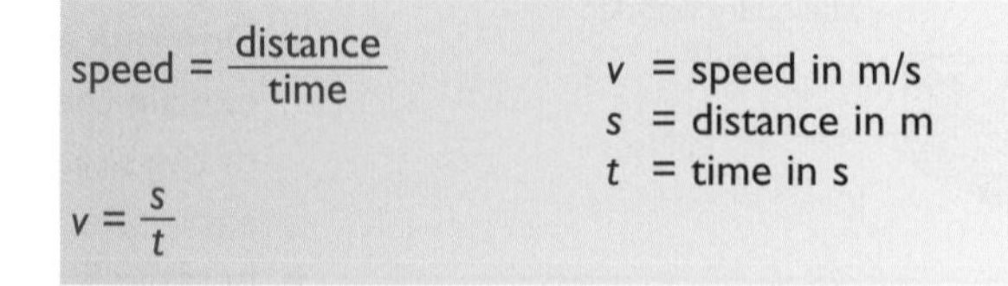

$$speed = \frac{distance}{time}$$

$$v = \frac{s}{t}$$

v = speed in m/s
s = distance in m
t = time in s

- **Velocity** is almost the same as speed. It has a **size** (called speed) and a **direction**.

WORKED EXAMPLES

1 Calculate the average speed of a motor car that travels 500 m in 20 seconds.

Write down the formula:	$v = \frac{s}{t}$
Substitute the values for s and t:	$v = \frac{500}{20}$
Work out the answer and write down the units:	v = 25 m/s

2 A horse canters at an average speed of 5 m/s for 2 minutes. Calculate the distance it travels.

Write down the formula in terms of s:	$s = v \times t$
Substitute the values for v and t:	$s = 5 \times 2 \times 60$
Work out the answer and write down the units:	s = 600 m

What is acceleration?

- How much an object's **speed or velocity changes** in a certain time is its **acceleration**. Acceleration can be calculated using the following formula:

$$acceleration = \frac{\text{change in speed}}{\text{time taken}}$$

$$a = \frac{(v - u)}{t}$$

a = acceleration
v = final speed in m/s
u = starting speed in m/s
t = time in s

- The units of acceleration m/s/s (metres per second per second) are sometimes written as m/s^2 (metres per second squared).

WORKED EXAMPLE

Calculate the acceleration of a car that travels from 0 m/s to 28 m/s in 10 seconds.

Write down the formula:	$a = \frac{(v - u)}{t}$
Substitute the values for v, u and t:	$a = \frac{(28 - 0)}{10}$
Work out the answer and write down the units:	$a = 2.8$ m/s/s

QUESTION SPOTTER

- Calculations of acceleration are very common. The equation will not be given in the exam so you will need to remember it. The questions usually will ask for a unit for the acceleration (e.g. *m/s/s*).

Vectors and scalars

- Velocity and acceleration are examples of **vector** quantities. A vector has a specific direction as well as a size, with a unit. Force is another example of a vector quantity.
- Speed is in example of a **scalar** quantity. A scalar quantity has size only, with a unit. Mass is another example of a scalar quantity.
- To add vectors, draw them to scale, joining them in turn 'head to tail'. The final, or **resultant**, vector is drawn from the 'tail' of the first vector to the 'head' of the last one.

A* EXTRA

- Displacement is the vector quantity linked to distance, which is a scalar quantity. Displacement is the distance travelled *in a particular direction.*

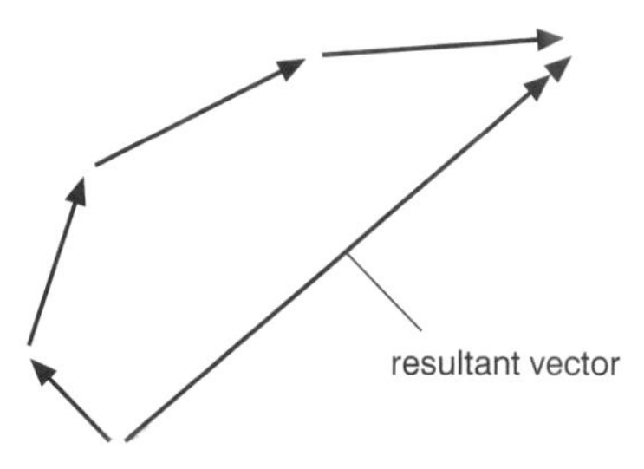

The resultant vector represents the combined effect of the individual vectors.

Using graphs to study motion

- Journeys can be summarised using **graphs**. The simplest type is a **distance-time graph** where the distance travelled is plotted against the time of the journey.

distance covered
0
time

distance covered
0
time

Steady speed is shown by a straight line. Steady acceleration is shown by a smooth curve.

A distance–time graph for a bicycle travelling down a hill. The graph slopes when the bicycle is moving. The slope gets steeper when the bicycle goes faster. The slope is straight (has a constant gradient) when the bicycle's speed is constant. The line is horizontal when the bicycle is at rest.

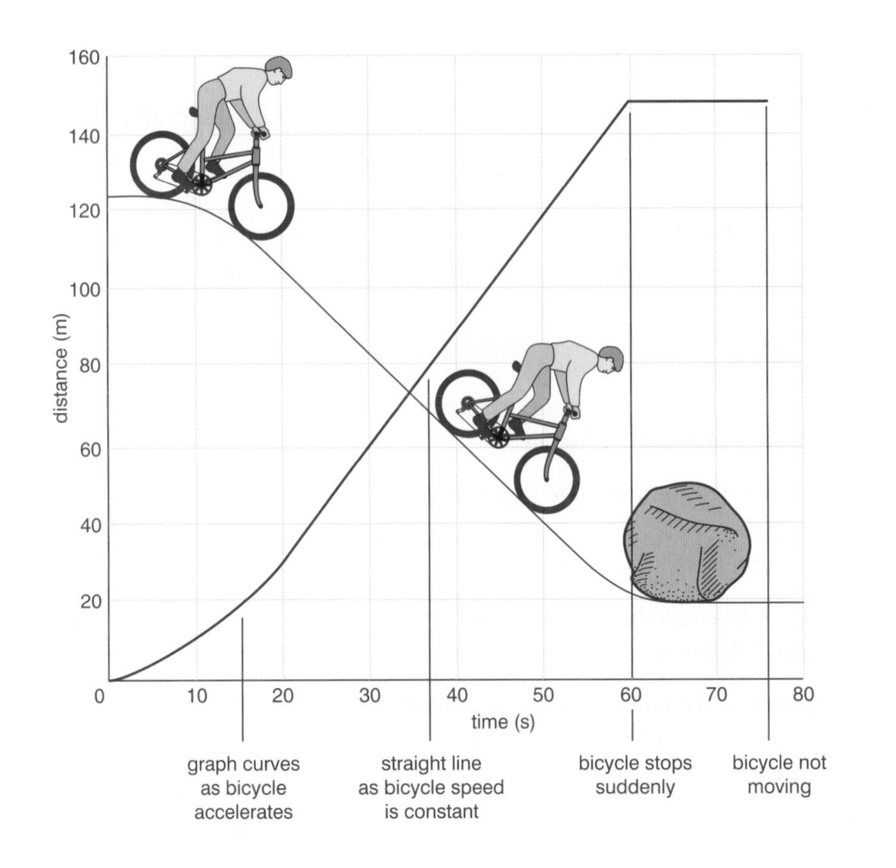

- A **speed-time graph** provides information on speed, acceleration and distance travelled.

Steady speed is shown by a horizontal line. Steady acceleration is shown by a line sloping up.

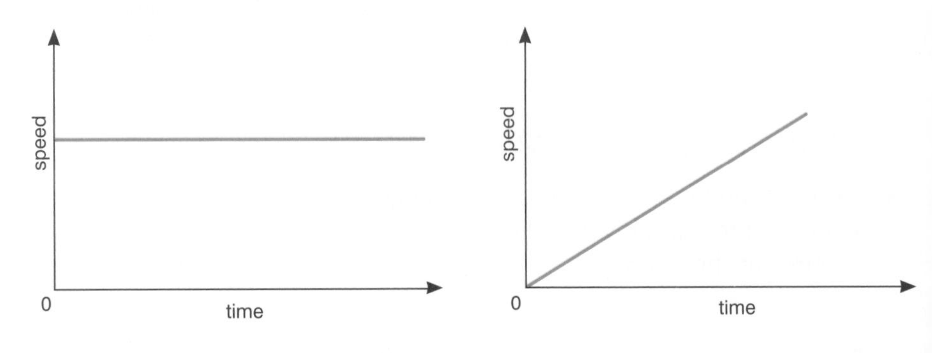

- The graph on the next page shows a car travelling between two sets of traffic lights. It can be divided into three regions.

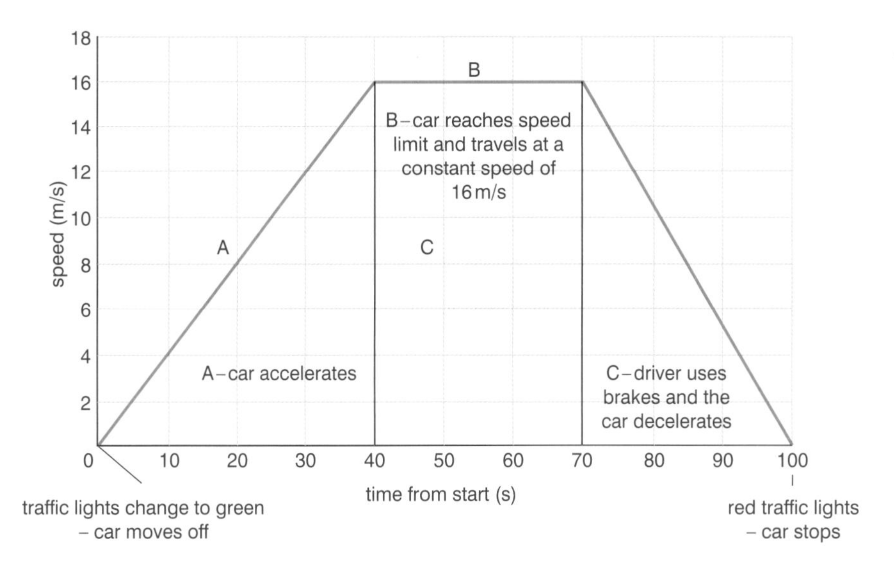

A speed–time graph for a car travelling between two sets of traffic lights.

- In region A, the car is **accelerating at a constant rate** (the line has a constant positive gradient). The distance travelled by the car can be calculated:

average velocity = $\frac{(16 + 0)}{2}$ = 8 m/s

time = 40 s

so distance = $v \times t$ = 8 × 40 = 320 m

This can also be calculated from the area under the line

($\frac{1}{2}$ base × height = $\frac{1}{2}$ × 40 × 16 = 320 m).

- In region B, the car is travelling at a **constant speed** (the line has a gradient of zero). The distance travelled by the car can be calculated:

velocity = 16 m/s

time = 30 s

so, distance = $v \times t$ = 16 × 30 = 480 m

This can also be calculated from the area under the line
(base × height = 30 × 16 = 480 m).

- In region C, the car is **decelerating at a constant rate** (the line has a constant negative gradient). The distance travelled by the car can be calculated:

average velocity = $\frac{(16 + 0)}{2}$ = 8 m/s

time = 30 s

so, distance = $v \times t$ = 8 × 30 = 240 m

This can also be calculated from the area under the line

($\frac{1}{2}$ base × height = $\frac{1}{2}$ × 30 × 16 = 240 m).

QUESTION SPOTTER

- You are likely to be asked to interpret distance/time and speed/time graphs. This will involve describing in words the movement of the object at different times during its journey. If the graph has a number of different regions then explain what is happening region by region.

Thinking, braking and stopping distances

- When a car driver has to brake, it takes time for him or her to react. During this time the car will be travelling at its normal speed. The distance it travels in this time is called the **thinking distance**.
- The driver then puts on the brakes. The distance the car travels while it is braking is called the **braking distance**. The overall **stopping distance** is made up from the thinking distance and the braking distance.
- The thinking distance can vary from person to person and from situation to situation – the braking distance can vary from car to car.

Factors affecting thinking distance	Factors affecting braking distance
speed	speed
tiredness	condition of tyres (amount of tread)
alcohol	condition of brakes
medication, drugs	road conditions (dry, wet, icy, gravel, etc.)
level of concentration and distraction	mass of the car

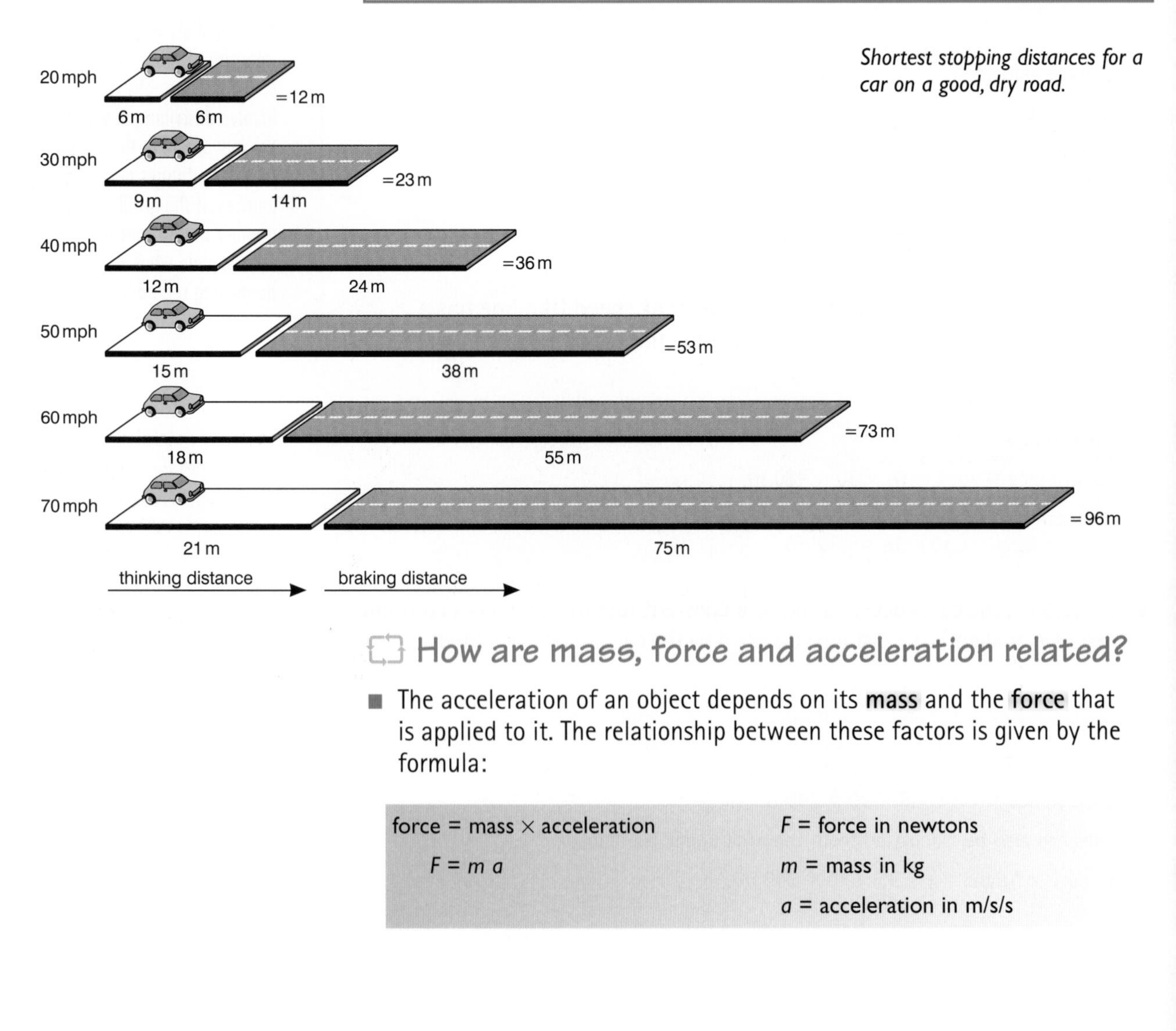

Shortest stopping distances for a car on a good, dry road.

How are mass, force and acceleration related?

- The acceleration of an object depends on its **mass** and the **force** that is applied to it. The relationship between these factors is given by the formula:

force = mass × acceleration

$F = m\,a$

F = force in newtons

m = mass in kg

a = acceleration in m/s/s

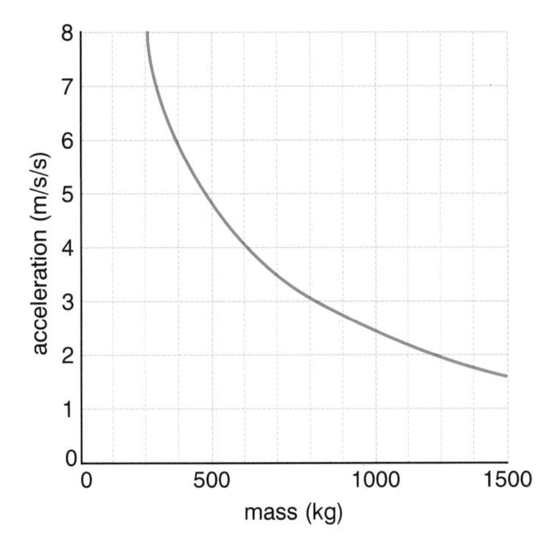

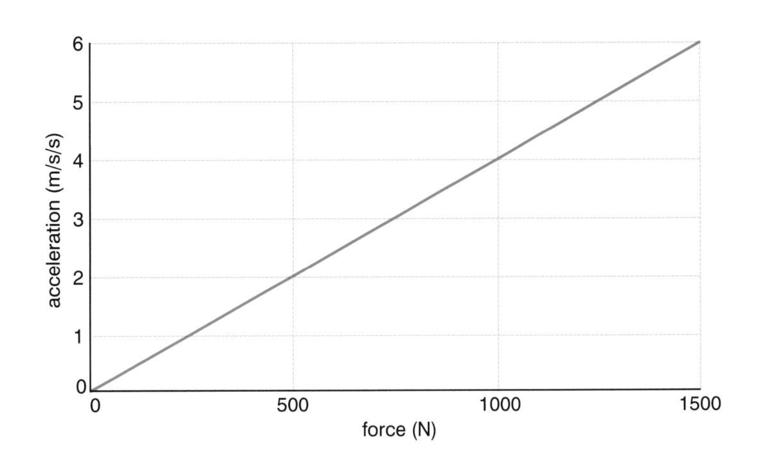

Acceleration is directly proportional to force. Acceleration is inversely proportional to mass.

WORKED EXAMPLES

1 What force would be required to give a mass of 5 kg an acceleration of 10 m/s/s?

Write down the formula:	$F = m\,a$
Substitute the values for m and a:	$F = 5 \times 10$
Work out the answer and write down the units:	$F = 50$ N

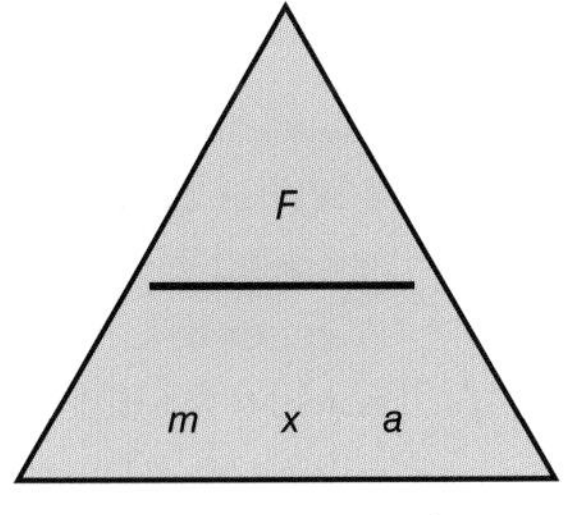

2 A car has a resultant driving force of 6000 N and a mass of 1200 kg. Calculate the car's initial acceleration.

Write down the formula in terms of *a*:	$a = \frac{F}{m}$
Substitute the values for *F* and *m*:	$a = \frac{6000}{1200}$
Work out the answer and write down the units:	$a = 5$ m/s/s

A* EXTRA

- The equation $F = m\,a$ shows that the acceleration of an object is directly proportional to the force acting (if its mass is constant) and is inversely proportional to its mass (if the force is constant). The gradient of a force–acceleration graph gives the mass of the object.

CHECK YOURSELF QUESTIONS

Q1 The graph shows a distance–time graph for a journey.

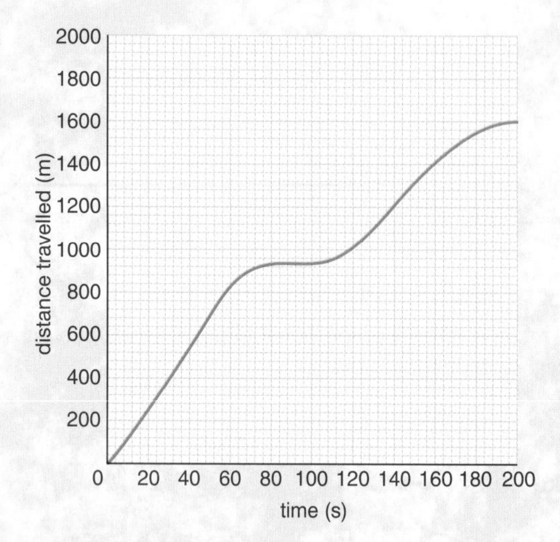

a What does the graph tell us about the speed of the car between 20 and 60 seconds?

b How far did the car travel between 20 and 60 seconds?

c Calculate the speed of the car between 20 and 60 seconds.

d What happened to the car between 80 and 100 seconds?

Q2 Look at the velocity–time graph for a toy tractor.

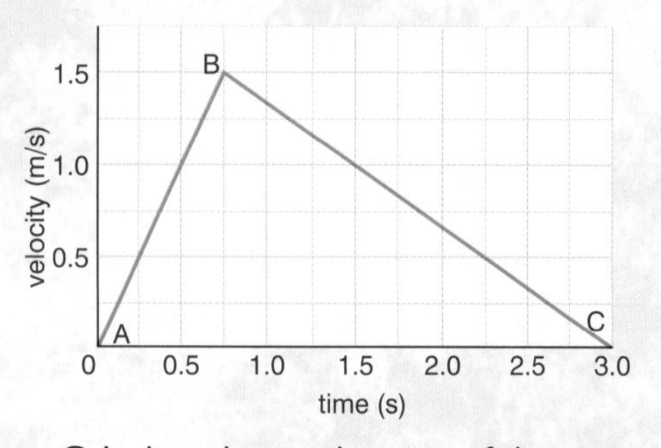

a Calculate the acceleration of the tractor from A to B.

b Calculate the total distance travelled by the tractor from A to C.

Q3 The manufacturer of a car gave the following information:
Mass of car 1000 kg. The car will accelerate from 0 to 30 m/s in 12 seconds.

a Calculate the average acceleration of the car during the 12 seconds.

b Calculate the force needed to produce this acceleration.

Q4 Peter cycles northwards for 30 seconds at 4 m/s. Then he cycles westwards for 30 seconds at 3 m/s.

a Calculate the distance Peter travels in each part of the journey.

b Use a vector diagram to calculate his overall displacement for the journey.

Answers are on page 140.

Vehicle safety features

Why do we need safety features?

- During **collisions**, cars and other vehicles stop in very short periods of time. Changing speed so quickly means that **very large accelerations** are involved, so **very large forces** are generated – remember $F = ma$. Larger forces cause **more serious injuries** to the passengers.
- Several safety features in vehicles work together to:
  - make the passenger slow down more gradually
  - which makes the acceleration smaller
  - which will make the forces smaller
  - which will reduce injuries.

How do crumple zones work?

crumple zone

The crumple zone helps the passenger cabin to slow down more gradually.

- Vehicles have crumple zones at the front and at the rear. These areas are **designed to crush** during an accident.
- Although the end of the vehicle stops very quickly, the passenger area stops more slowly. This reduces the acceleration of the passengers, which reduces the forces on them.

How do seat belts work?

- When the vehicle stops, passengers tend to keep going until something stops them. Without a seat belt, the passengers may stop very suddenly when they hit the windscreen or a passenger in front of them.

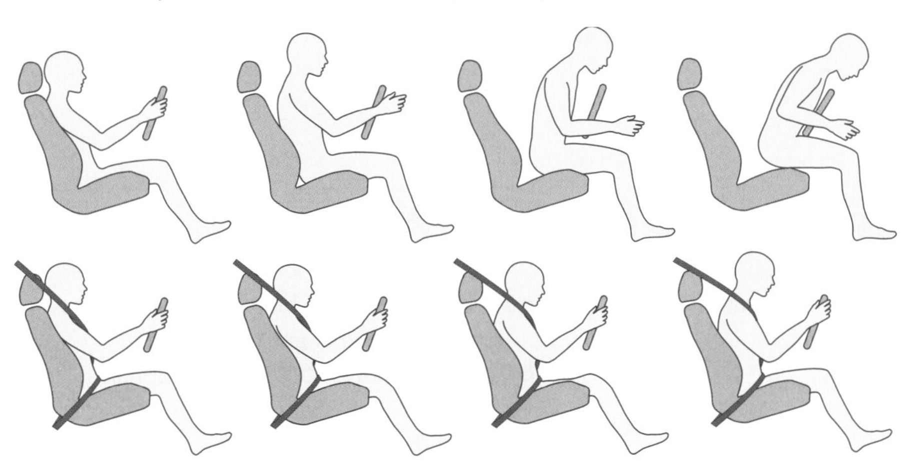

Seat belts help save many injuries by holding passengers in their seats.

- Seat belts are designed to hold passengers in the seat so that they stop **with** the vehicle. This reduces their acceleration, which reduces the forces on them.
- Seat belts are designed to stretch a little during a crash. If they did not, they would hold the person in place too strongly, which would make the person stop too quickly and increase the forces. Because of this stretching, seat belts should be replaced after a collision.

- Remember to mention that large forces cause more serious injuries.

A* EXTRA

- Avoid saying that passengers are 'thrown forward' in a crash. Although that is what it feels like, remember that the passenger is just continuing their motion forward – the car has stopped beneath them.

How do airbags work?

- An air bag inflates quickly during a collision and then immediately begins to deflate. Without the airbag, the passenger will tend to keep moving forward until they stop quickly against a hard object such as the steering wheel.
- As the air bag deflates, the passenger **slows down slowly**, reducing the acceleration and hence, the forces.

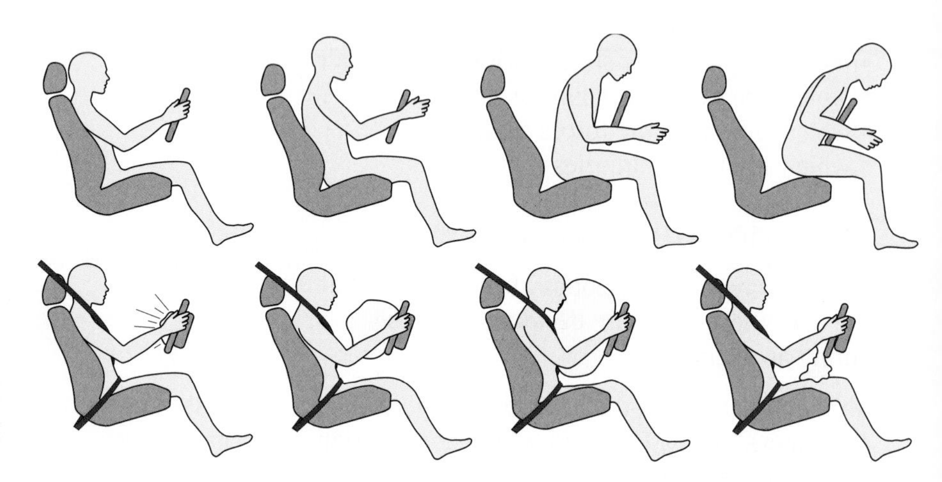

An air bag also slows the passenger gradually, reducing the risk of injury.

QUESTION SPOTTER

- It is common to use this section to check on your quality of written communication. You should be able to link several sentences together to describe how these safety features work.

IDEAS AND EVIDENCE

Air bags are designed to work against the chest of the passenger. For small children and babies in car seats this is a real problem – the air bag itself can cause major injuries in these cases. Be prepared to give points for and against their use.

CHECK YOURSELF QUESTIONS

Q1 Explain why it is important that a seat belt stretches a little during a crash.

Q2 **a** An air bag may not help at all during a very high speed crash. Explain why.

b An air bag starts to deflate as soon as it has blown up. Explain how this helps reduce the forces involved.

Answers are on page 141.

Using the equations of motion

How can we describe the motion of an object?

- If the motion of an object has a **constant acceleration**, then there are three more formulas that can be useful.

$$v = u + at$$
$$v^2 = u^2 + 2as$$
$$s = ut + \frac{1}{2}at^2$$

s = distance travelled in m
u = starting velocity in m/s
v = final velocity in m/s
a = acceleration in m/s^2
t = time in s

QUESTION SPOTTER

- These formulas will be given to you on the exam paper, but you might have to re-arrange them.

- Follow a routine when answering questions using the formulas:
  - Write down the letters s, u, v, a, t.
  - Beside them, write down the information given in the question.
  - Decide which quantity you need to calculate.
  - Choose the formula that you need.
  - Re-arrange the formula if necessary.
  - Substitute the values, changing the units if necessary.
  - Calculate your answer.

WORKED EXAMPLE

When an aircraft take off, it accelerates from 0 m/s to 90 m/s on a runway that is 1500 m long. Find the acceleration.

Write down the letters and the information given:
s = 1500 m
u = 0 m/s
v = 90 m/s
a = calculate this
t = no information

Choose a formula that does not involve t: $v^2 = u^2 + 2as$

Re-arrange to give a: $a = \frac{v^2 - u^2}{2s}$

Substitute the values for s, u and v: $a = \frac{90^2 - 0^2}{2 \times 1500}$

Work out the answer and write down the units: $a = 2.7$ m/s^2

- Make sure you learn the unit of acceleration.

Code words in the question

- Sometimes the information is given in the text, rather than as numbers. Watch out for:
  - '**Starts from rest**' tells you the **initial** velocity is 0 m/s.
  - '**Comes to a stop**' tells you the **final** velocity is 0 m/s.

WORKED EXAMPLE

A car is travelling at 30 m/s. The driver applies the brakes and comes to a stop in 25 s. Work out the acceleration.

Write down the letters and the information given:
s = no information
u = 30 m/s
v = 0 m/s ('comes to a stop')
a = calculate this
t = 25 s

Choose a formula that does not involve s: $v = u + at$

Re-arrange to give a: $a = \frac{v - u}{t}$

$a = \frac{0 - 30}{25}$

Work out the answer and write down the units: $a = -1.2$ m/s^2

The minus sign tells us that the car is slowing down – it is a deceleration.

When gravity is involved

- Questions usually involve gravity in one of two ways:
  - An object **falls** from a height, for example a stone falling from a cliff.
  - An object is **launched** upwards and then falls back down.
- When answering these questions remember that the acceleration is caused by gravity and is **always** directed **vertically downwards**.
  - The value of the acceleration, g, will be given as 10 m/s^2.

WORKED EXAMPLES

1 A stone is dropped from a cliff onto the beach beneath. It lands on the beach after falling for 3 seconds. Find the height of the cliff.

Write down the letters and the information given:
s = calculate this
u = 0 m/s ('dropped')
v = no information
a = acceleration due to gravity (g = 10 m/s^2)
t = 3 s

Choose a formula that does not involve v: $s = ut + \frac{1}{2}at^2$

$s = (0 \times 3) + \frac{1}{2} \times 10 \times 3^2$

Work out the answer and write down the units: s = 45 m

2 Sally throws a ball upwards at 10 m/s. How high does she throw the ball?

The important thing is to treat the motion as **two separate parts** – the ball going up (and stopping) and then falling back down again.

Thinking **only** about the ball moving from her hand to the top of its flight:

Write down the letters and the information given:
s = calculate this
u = 10 m/s
v = 0 m/s (at the top of the flight the ball stops before it falls back)
a = acceleration due to gravity ($g = -10$ m/s^2 is negative because the motion is upwards, but the gravity is decelerating the ball, pulling it backwards)
t = no information

Choose the formula that uses s, v, u and a: $v^2 = u^2 + 2as$

Re-arrange to give s: $s = \frac{v^2 - u^2}{2a}$

$s = \frac{0^2 - 10^2}{2 \times -10}$

Work out the answer and write down the units: $s = 5$ m

- If you need to find a force in these questions, then you will probably need to make use of $F = ma$. Information about the mass should be given to you.

CHECK YOURSELF QUESTIONS

Q1 A train sets off from a station and accelerates at 0.2 m/s^2 for 2 minutes. How far will it travel?

Q2 A Formula 1 racing car accelerates from rest to 54 m/s in 3 s. Calculate the acceleration.

Q3 During a collision, a car slows from 15 m/s to 0 m/s in 10 m. Find the deceleration.

Answers are on page 141.

Motion in two dimensions

What is the effect of gravity?

As an object falls, it accelerates downwards due to the force of gravity.

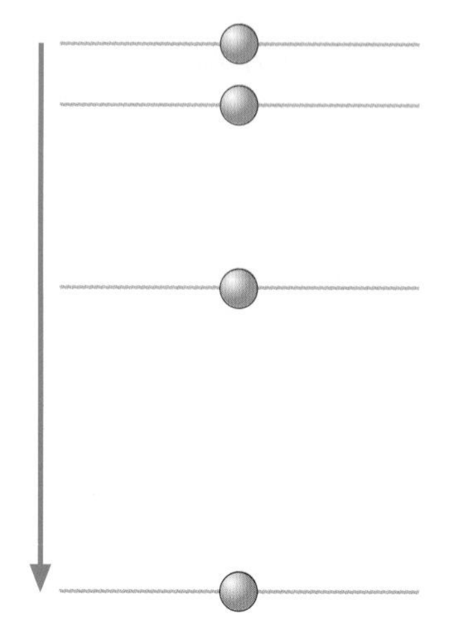

- Gravity is a force that controls the motion of stars, galaxies, planets and all other objects in the universe. When dealing with motion close to the Earth's surface, however, the rule is quite simple – gravity always acts **towards** the centre of the Earth. For most situations this means **vertically downwards**.
- The force of gravity causes objects to accelerate downwards. If we ignore air resistance and any other factors that might work against gravity, then an object will continue to accelerate at a rate of 10 m/s^2 until it hits the Earth.

How does a sideways force cause a projectile to move?

An object travelling horizontally, without any effect of gravity, has constant speed.

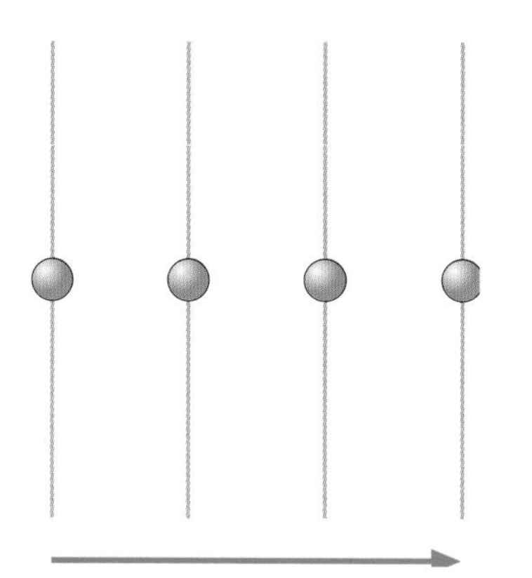

- A projectile is an object that only has a force acting on it at launch. Once the object is in motion, the object only changes speed if another force starts to act on it. Examples include throwing a tennis ball, firing a cannon, playing darts and so on.
- If we ignore air resistance and any effect of gravity, then an object projected horizontally will continue to move at a constant speed.

What about the two effects combined?

The stone follows the curved path of a parabola as it has both horizontal and vertical speeds.

- Think about a stone thrown sideways from a cliff. Ignoring air resistance, we can treat the horizontal and vertical parts of the motion completely separately.
- **Horizontally**, the object has a constant speed.
- **Vertically**, gravity causes the object to accelerate downwards at 10 m/s^2.
- The combination gives a curved path called a **parabola**.
- The sideways motion does *not* change the effect of gravity – objects thrown sideways land at the same time as objects that fall straight down.

- If an object is fired upwards at an angle, for example a rugby ball kicked towards a goal, then the parabola shape is repeated. As the ball moves **upwards**, it **decelerates** due to gravity until it reaches the highest point, and then the ball **accelerates** as it comes **down** again. The horizontal velocity stays the same throughout.

QUESTION SPOTTER

- Questions often ask you to draw in the path of the object. Make sure your line starts to curve downwards as soon as the object is released.

A* EXTRA

- A parabola is different from a circle. Because the object accelerates due to gravity, the path becomes increasingly steep as the object falls.

IDEAS AND EVIDENCE

The effect of gravity on motion has been studied for a long time. According to legend, Galileo dropped a heavy object and a light object from the leaning tower of Pisa to see which would fall faster. They landed together – although gravity does pull a heavier object with a greater force, its extra mass means that the acceleration always comes out the same. Check $F = m\,a$ in Revision Session 2.

CHECK YOURSELF QUESTIONS

Q1 Danny is at the fairground on the coconut shy. He aims a ball straight at a coconut. The ball misses.

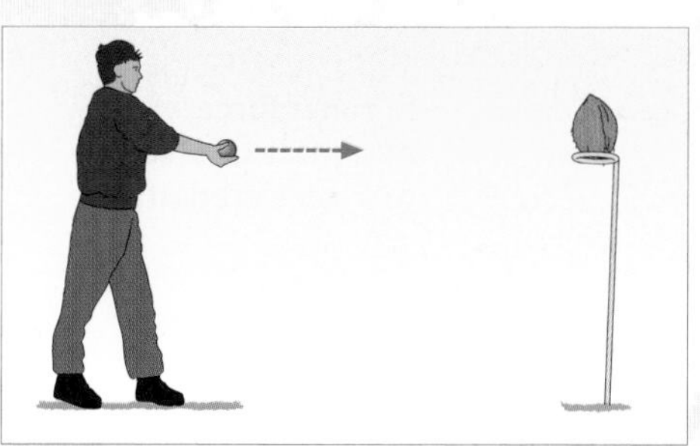

a Copy the diagram and draw the path of the ball.

b Draw a second line to show how Danny should have aimed the ball.

Q2 Raj experiments by releasing two tennis balls from an upstairs window. He throws ball A sideways at 5 m/s and allows ball B to fall. Raj lets go of both balls at the same time.

a Which ball (if any) lands first? Explain your answer.

b If ball A lands 20 m away from the building, calculate the time it takes from release to landing.

Answers are on page 141.

Momentum

What is momentum?

- **Momentum** is a useful idea that helps us analyse the motion of objects.
- Momentum can be calculated using the formula:

momentum = mass × velocity
momentum = $m\,v$

m = mass in kg
v = velocity in m/s
Momentum is measured in kg m/s.

QUESTION SPOTTER

- The unit of momentum is quite a tricky one – usually kg m/s. It will often be given to you in the question but make sure you learn it – if it isn't given in the answer space there will be a mark for it.

WORKED EXAMPLE

Calculate the momentum of a racing car of mass 600 kg travelling at 75 m/s.

Write down the formula:	momentum = mass × velocity
Substitute the values for m and v:	momentum = 600 kg × 75 m/s
Work out the answer and write down the units:	momentum = 45 000 kg m/s

Changing momentum

- Velocity is a vector quantity – it has a **direction** – so **momentum has a direction**. When working out problems, one direction is called the **positive** direction (for example, going to the right), with the opposite direction taken as the **negative** direction (for example, going to the left).
- The momentum of an object will change if:
  - the mass changes
  - the velocity changes
  - the direction of the movement changes.
- The momentum of an object changes when a **force** acts on it.

Momentum and Newton's second law of motion

- For GCSE Science, Newton's second law of motion is usually given as:

net force = mass × acceleration
$F = ma$

F = net force
m = mass of object
a = acceleration

- A fuller version of this law is:

net force = rate of change of momentum

$$F = \frac{mv - mu}{t}$$

v = final velocity
u = starting, or initial, velocity
t = time for which the force is acting

- This version works **provided the mass of the object does not change.**

A* EXTRA

- There are a few situations where the mass does change. For example, when a rocket launches the fuel is burned and so the overall mass of the rocket decreases. In situations like this, use two separate values for the mass when you work out the starting momentum and final momentum.

- Re-arranging the formula gives:

net force = rate of change of momentum

$$F = \frac{m\,(v - u)}{t}$$

which leads back to $F = ma$, since acceleration, $a = \frac{(v - u)}{t}$

- Re-arranging the formula in a slightly different way gives:

$$Ft = mv - mu$$

net force × time the force is acting = change in momentum

- This formula helps to illustrate two important ideas.
  - For a particular force, there is a **greater** change in momentum if the force acts for a **longer** time. For example, a footballer who follows through when kicking, makes contact with the ball for longer, so there is a greater change in the ball's momentum and the ball flies faster.
  - For a particular *change in momentum*, the longer the change takes, the **smaller** the force will be. For example, when a parachutist lands, he bends his knees. This makes the momentum change take a longer time, so the force on the parachutist is smaller. This is also very important in **car safety features** such as **crumple zones** – the car is designed to crumple on impact which makes the momentum change take place over a longer time, reducing the forces on the occupants.

Conservation of momentum

- In any collision or explosion, the total momentum after the event is equal to the total momentum before the event, as long as no external forces are acting on the system.
- This is the **principle of conservation of momentum**.
- In problems using this principle, work through these steps:
  - Draw two diagrams of the situation, labelled 'before' and 'after'.
  - Mark the values given in the question, remembering to use '+' and '–' to show directions if necessary.
  - State the principle.
  - Substitute the values from the question.
  - Work out the answer, adding the unit if necessary.

WORKED EXAMPLES

1 In an experiment, a 2 kg trolley, travelling at 0.5 m/s collides with, and sticks to, a 3 kg trolley that is initially stationary. Work out the velocity of the pair after the collision.

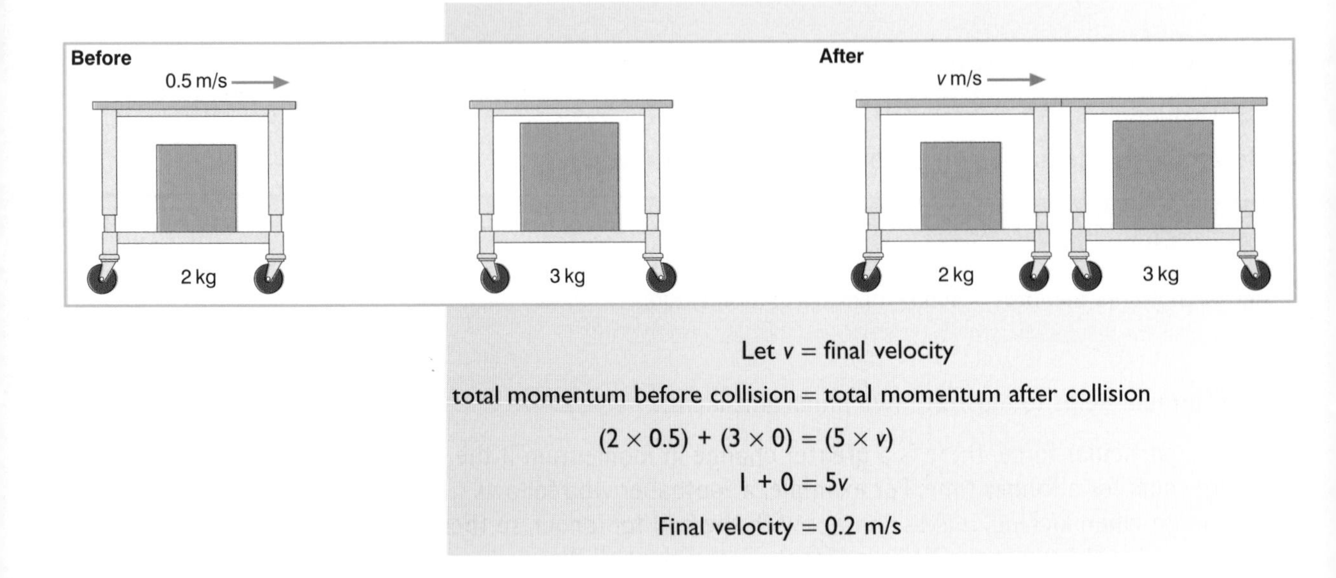

Let v = final velocity

total momentum before collision = total momentum after collision

$$(2 \times 0.5) + (3 \times 0) = (5 \times v)$$

$$1 + 0 = 5v$$

Final velocity = 0.2 m/s

2 Two cars collide head-on as shown in the diagram. After the collision both cars are stationary. Calculate the velocity of the 750 kg car before the collision.

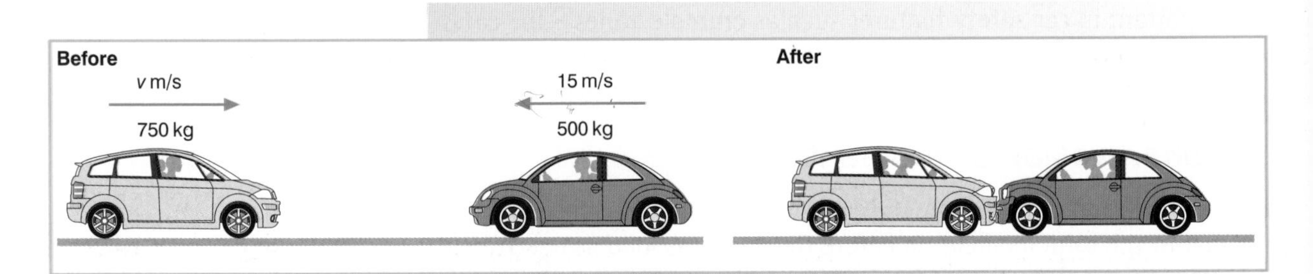

Let v = required velocity

total momentum before collision = total momentum after collision

$$(750 \times v) + (500 \times -15) = (750 \times 0) + (500 \times 0)$$

$$750v - 7500 = 0$$

$$750v = 7500$$

$$v = 10 \text{ m/s}$$

3 When a rifle fires, the bullet is projected forwards at a speed of 400 m/s. The rifle moves backwards (it recoils) at a speed of 0.4 m/s. If the rifle has a mass of 10 kg, find the mass of the bullet.

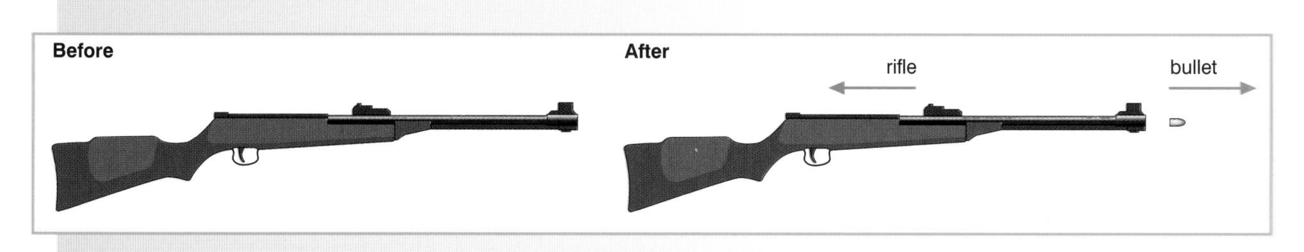

Let m = mass of bullet

total momentum before = total momentum after

$$(10 \times 0) + (m \times 0) = (10 \times -0.4) + (400 \times m)$$

$$0 = -4 + 400m$$

$$4 = 400m$$

mass of bullet, $m = 0.01$ kg (i.e. 10 g)

CHECK YOURSELF QUESTIONS

Q1 A car of mass 500 kg accelerates from 15 m/s to 30 m/s.

a Calculate the change in momentum of the car.

b If the car takes 5 s to make this change, calculate the net force required.

c The actual force provided by the car engine will be larger than the value calculated in part **b**. Explain why.

Q2 Two cars collide head on as shown in the diagram. The cars stick together in the collision and move off together afterwards.

a Calculate the momentum of each car before the collision. (Remember to use '+' and '–')

b Calculate the total momentum before the collision.

c Write down the total momentum after the collision.

d Find the velocity of the cars after the collision. In which direction do they move?

Q3 David is standing on a stationary skateboard. When he steps off the skateboard forwards the skateboard moves backwards. Use the principle of conservation of momentum to explain why this happens.

Answers are on page 141.

REVISION SESSION 7

Turning forces

The turning effect of a force

- The turning effect of a force is called the **moment** of the force.
- The moment of a force depends on two things:
  - the size of the force
  - the distance between the line of the force and the turning point, which is called the **pivot**.
- We calculate the moment of force using this formula:

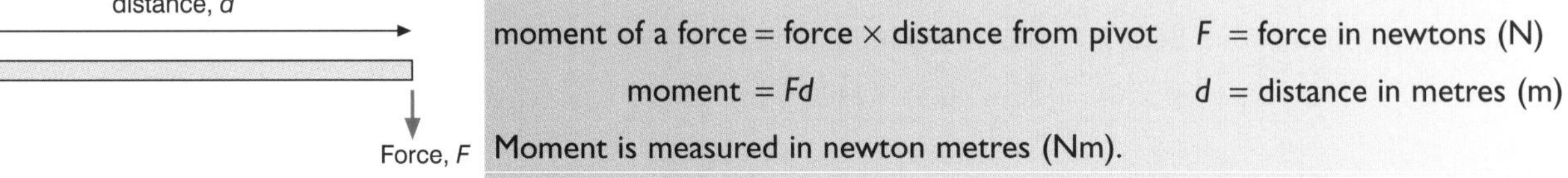

moment of a force = force × distance from pivot F = force in newtons (N)

moment = Fd d = distance in metres (m)

Moment is measured in newton metres (Nm).

QUESTION SPOTTER

- Learn the unit for a moment, Nm, or sometimes Ncm, if the distance is in cm.

WORKED EXAMPLE

Michelle pushes open a door with a force of 20 N. The door is 0.8 m wide. Calculate the moment of this force.

Write down the formula:	moment = force × distance from pivot
Substitute the values for F and d:	moment = 20 N × 0.8 m
Work out the answer and write down the units:	moment = 16 Nm

When objects balance

- If an object is not turning, **the sum of the clockwise moments equals the sum of the anticlockwise moments**. This is the **principle of moments**.

WORKED EXAMPLE

Phil and Jenny are sitting on a see-saw. The see-saw is balanced. Work out Phil's weight.

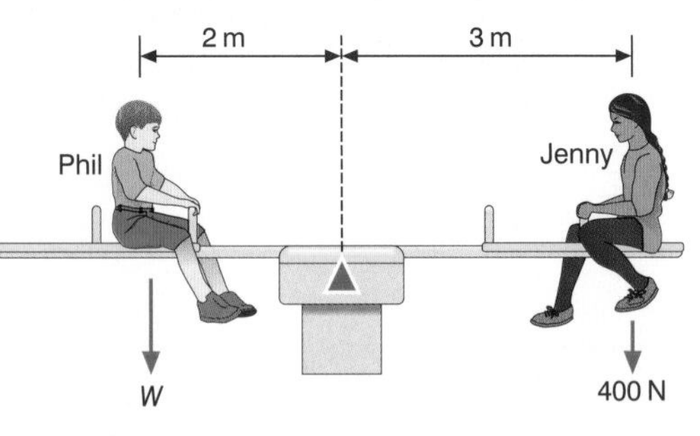

Jenny is causing the clockwise moment of 400 N × 3 m.

Phil is causing the anticlockwise moment of W × 2 m.

The see-saw is balanced, so

the sum of the clockwise moments = the sum of the anticlockwise moments

$$400 \times 3 = W \times 2$$

$$W = 600 \text{ N}$$

What is the centre of mass?

- The centre of mass is the point where we can assume *all* the mass of the object is concentrated. This is a useful simplification because we can pretend **gravity** only acts at a **single point** in the object, so a single arrow on a diagram can represent the **weight** of an object.
- The centre of mass for objects with a regular shape is in the centre.

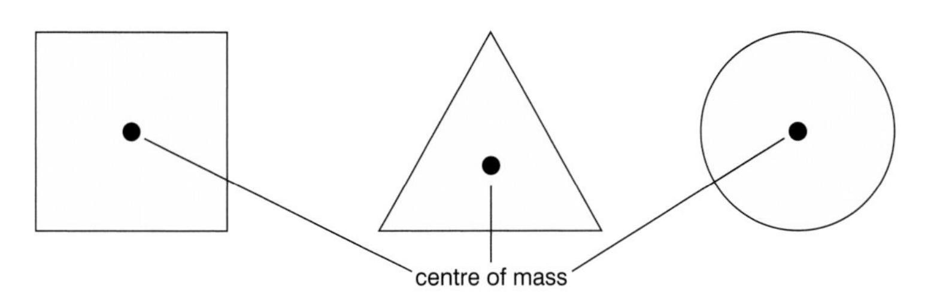

- Questions involving balancing can look quite difficult. Work out each moment in turn and add together the moments that turn the same direction.

What about irregular shapes?

- To find the centre of mass of simple objects, such as a piece of card, follow these steps:

1 Hang up the object.
2 Suspend a mass from the same place.
3 Mark the position of the thread.
4 The centre of mass is somewhere along the line of the thread.
5 Repeat steps 1 to 3 with the object suspended from a different place.
6 The centre of mass is where the two lines meet.

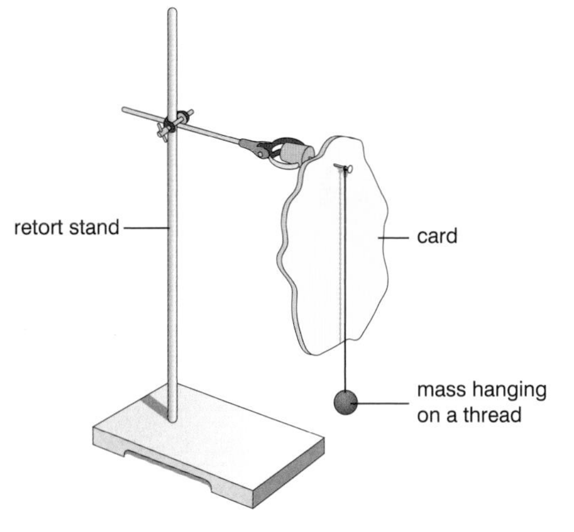

Centre of mass links to stability

- The idea of centre of mass is useful when predicting whether or not an object will fall over – whether or not it is **stable**.

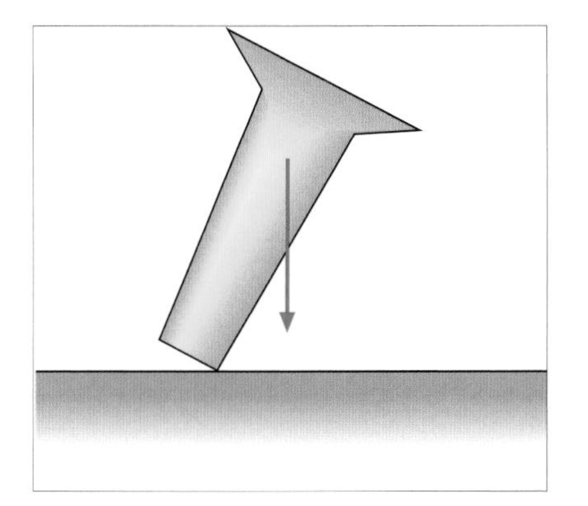

This object will topple over – the centre of mass is outside the pivot so the weight of the object tips it over the rest of the way. The moment of the force turns the object over. An object that is easy to topple is said to be in ***unstable equilibrium***.

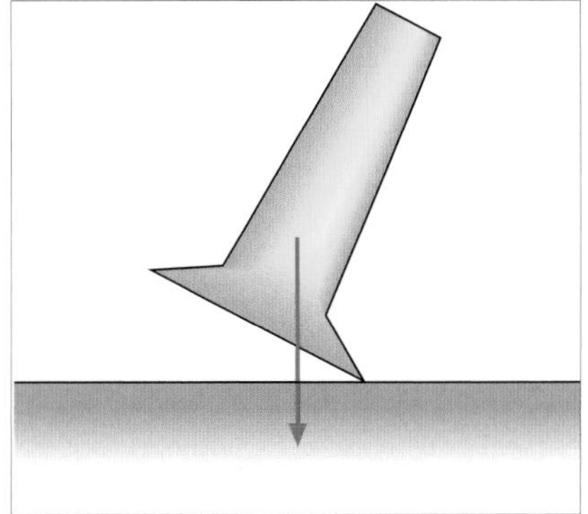

This object will fall back into place – the centre of mass is inside the pivot so the weight of the object pulls it back onto its base. The moment of the force returns the object to its base. An object that is difficult to topple is said to be in ***stable equilibrium***.

CHECK YOURSELF QUESTIONS

Q1 A flag is being blown by the wind. The force on the flag is 100 N and the flagpole is 8 m tall.

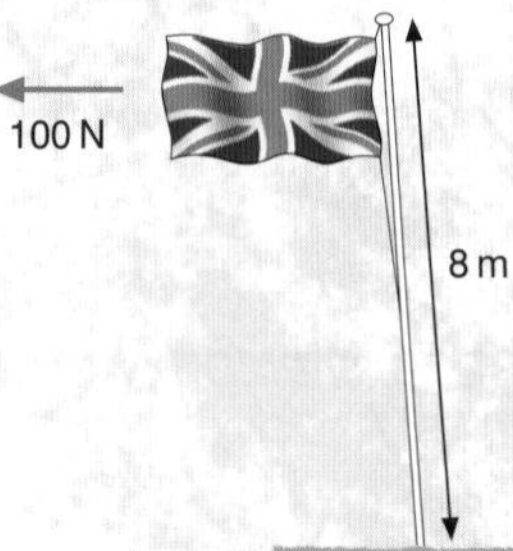

Calculate the moment of the force about the base of the flagpole.

Q2 Which of these glasses is the most stable? Explain your answer.

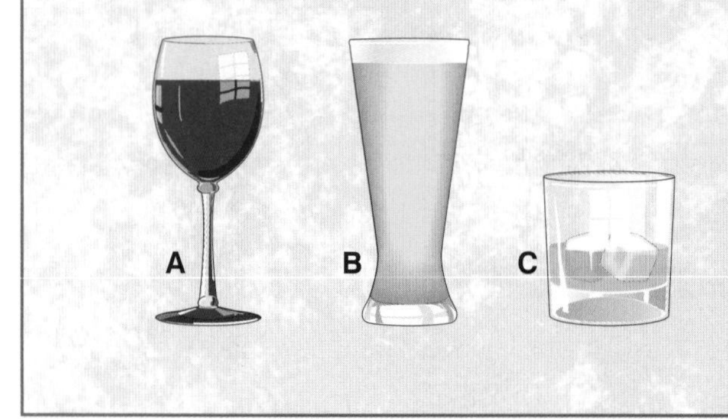

Q3 Rod and Jane are sitting on a see-saw. The see-saw is not balanced.

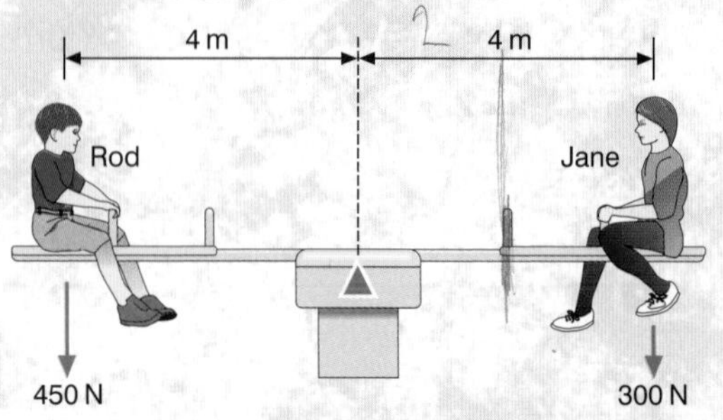

Freddy weighs 300 N. Where should Freddy sit in order to balance the see-saw?

Answers are on page 142.

Where does our energy come from?

Sources of energy

- Most of the energy we use is obtained from **fossil fuels** – coal, oil and natural gas.
- Once supplies of these fuels have been used up. they cannot be replaced – they are **non-renewable.**
- At current levels of use, oil and gas supplies will last for about another 40 years, and coal supplies for about a further 300 years. The development of **renewable** sources of energy is therefore becoming increasingly important.

Renewable energy sources

- **Solar power** is energy from the Sun. The Sun's energy is trapped by solar panels and transferred into electrical energy or, as with domestic solar panels, is used to heat water. The cost of installing solar panels is high, and the weather limits the time when the panels are effective.
- The **wind** is used to turn windmill-like turbines which generate electricity directly from the rotating motion of their blades. Modern wind turbines are very efficient but several thousand would be required to equal the generating capacity of a modern fossil-fuel power station.
- The motion of **waves** can be used to move large floats and generate electricity. A very large number of floats are needed to produce a significant amount of electricity.
- Dams on tidal estuaries trap the water at high tide. When the water is allowed to flow back at low tide, **tidal power** can be generated. This obviously limits the use of the estuary.

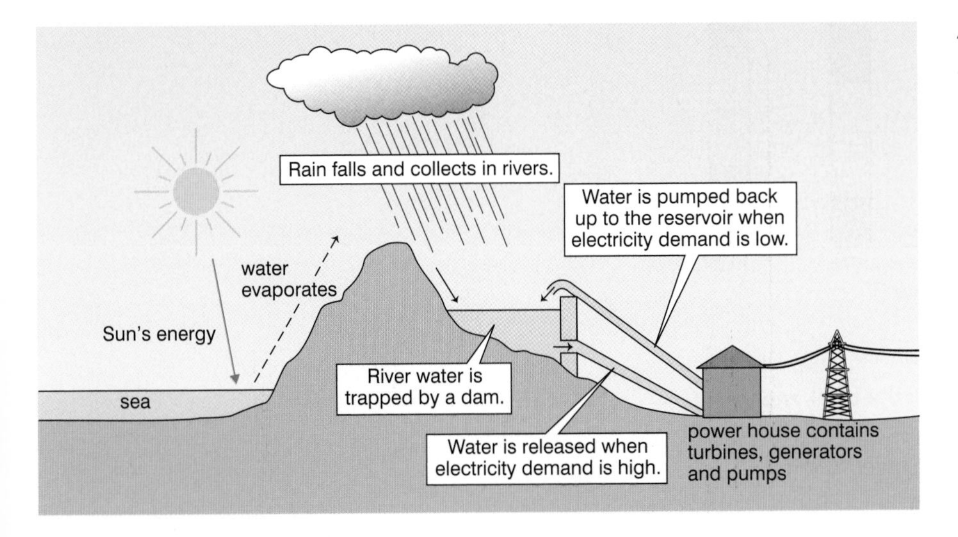

A 'pumped storage' hydroelectric power station.

IDEAS AND EVIDENCE

For a particular energy source, you should be able to give reasons *for* and *against* its use. For example, in favour of wind power, you might say that it is renewable and cheap to run, whilst you could say that wind farms cause visual pollution and are expensive to set up. Remember that there are reasons in favour of fossil fuels, particularly that they store large amounts of energy, as well as disadvantages such as the release of greenhouse gases when they are burned.

- Dams can be used to store **water** which is allowed to fall in a controlled way that generates electricity. This is particularly useful in mountainous regions for generating **hydroelectric power**. When demand for electricity is low, electricity can be used to pump water back up into the high dam for use in times of high demand.
- **Plants** use energy from the Sun in photosynthesis. Plant material can then be used as a **biomass fuel** – either directly by burning it or indirectly. A good example of indirect use is to ferment sugar cane to make ethanol, which is then used as an alternative to petrol. Waste plant material can be used in 'biodigesters' to produce methane gas. The methane is then used as a fuel.
- **Geothermal power** is obtained using the heat of the Earth. In certain parts of the world, water forms hot springs which can be used directly for heating. Water can also be pumped deep into the ground to be heated.

CHECK YOURSELF QUESTIONS

Q1 **a** What is meant by a non-renewable energy source?

b Name three non-renewable energy sources.

c Which non-renewable energy source is likely to last the longest?

Q2 Look at the graph, which shows the amount of energy from different sources used in the UK in 1955, 1965, 1975 and 1985.

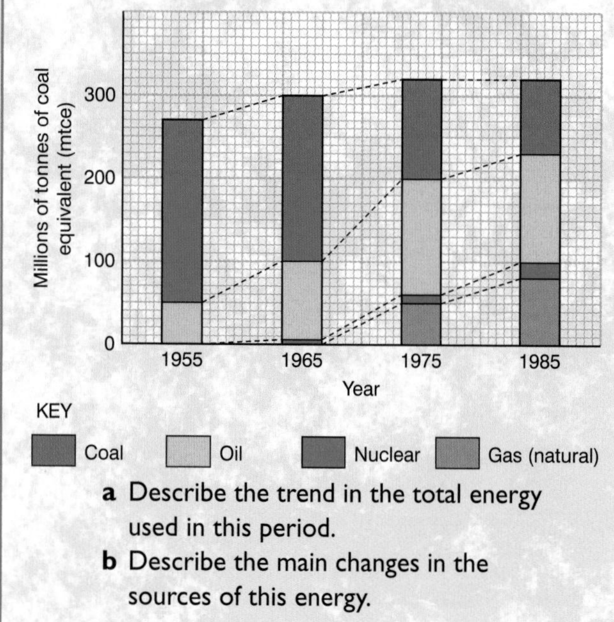

a Describe the trend in the total energy used in this period.

b Describe the main changes in the sources of this energy.

Q3 A site has been chosen for a wind farm (a series of windmill-like turbines).

a Give two important factors in choosing the site.

b Give one advantage and one disadvantage of using wind farms to generate electricity.

Answers are on page 142.

Transferring energy

How is energy transferred?

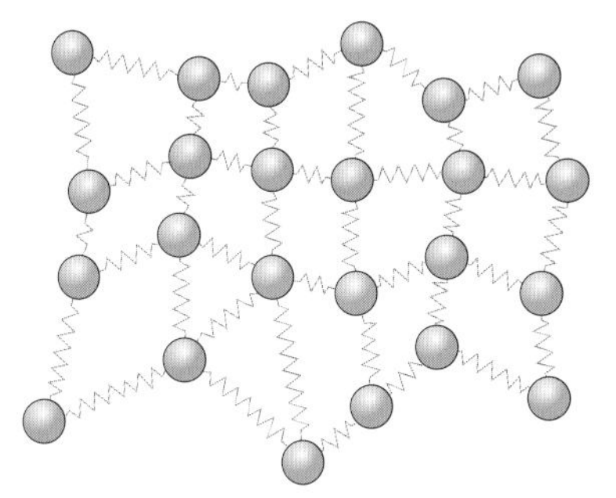

- Energy flows from high temperatures to low temperatures – this is called **thermal transfer**. Thermal energy can be transferred in four main ways:
  - **conduction**
  - **convection**
  - **radiation**
  - **evaporation.**

Conduction

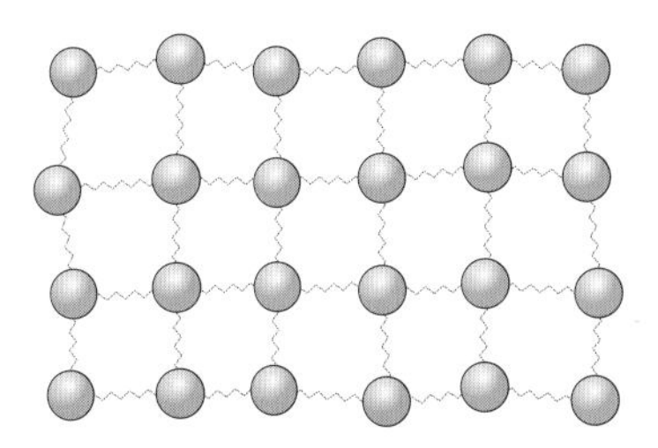

Conduction in a solid. Particles in a hot part of a solid (top) vibrate further and faster than particles in a cold part (bottom). The vibrations are passed on through the bonds from particle to particle.

- Materials that allow thermal energy to transfer through them quickly are called **conductors.** Those that do not are called **insulators.**
- If one end of a conductor is heated, the atoms that make up its structure start to vibrate more vigorously. As the atoms in a solid are linked together by chemical bonds, the increased vibration can be passed on to other atoms. The energy of movement (kinetic energy) passes through the whole material.
- **Metals** are particularly good conductors because they contain freely moving electrons which transfer energy very rapidly. **Air** is a good insulator. As air is a gas, there are no bonds between the particles and so energy can only be transferred by the particles colliding with each other. Conduction cannot occur when there are no particles present, so a vacuum is a perfect insulator.

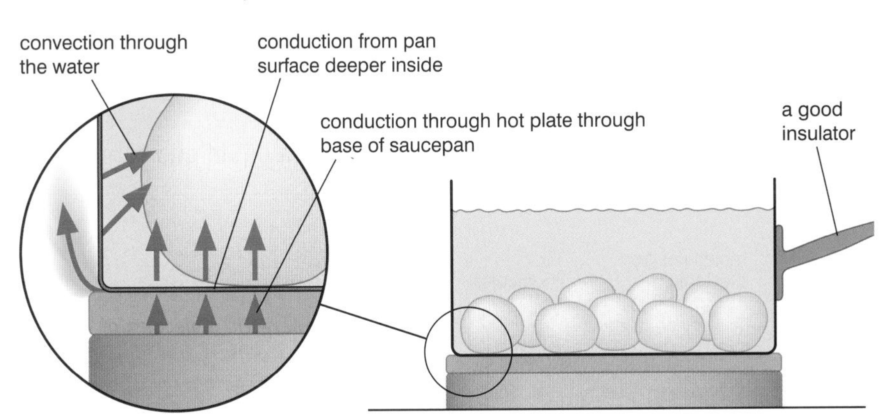

Conduction plays an important part in cooking food.

Convection

- Convection occurs in **liquids** and **gases** because these materials flow (they are 'fluids'). The particles in a fluid move all the time. When a fluid is heated, energy is transferred to the particles, causing them to move faster and further apart. This makes the heated fluid less dense than the unheated fluid. The less dense warm fluid will rise above the more dense colder fluid, causing the fluid to circulate. This **convection current** is how the thermal energy is transferred.

QUESTION SPOTTER

- Examination questions often ask for explanations of processes that combine conduction and convection – for example, in explaining the energy transfer in cooking potatoes (see diagram).

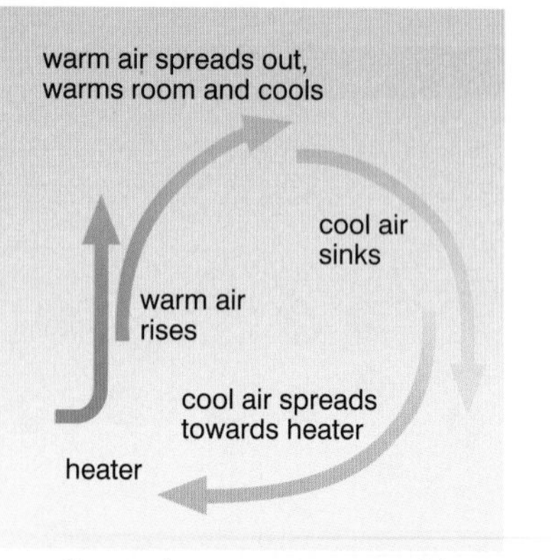

Convection currents caused by a room heater.

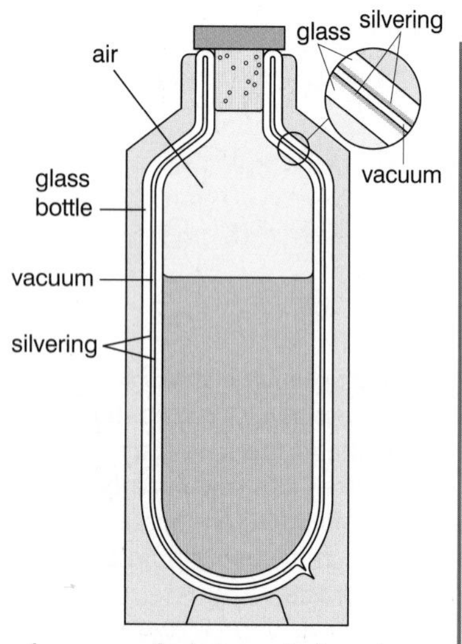

A vacuum flask has a hollow glass lining. The absence of air in the lining prevents thermal transfer by conduction and convection. The insides of the lining are silvered to reduce thermal transfer by radiation. The flask will therefore keep hot drinks hot, or cold drinks cold, for hours.

A* EXTRA

Convection currents in gases and liquids are caused by changes in density.

- When a liquid is heated the particles move more quickly and move further apart, making that part of the liquid less dense than cooler parts.
- The less dense part then rises and the cooler or denser liquid takes its place.

- If a fluid's movement is restricted, then energy cannot be transferred. That is why many insulators, such as ceiling tiles, contain trapped air pockets. Wall cavities in houses are filled with fibre to prevent air from circulating and transferring thermal energy by convection.

Radiation

- Radiation, unlike conduction and convection, **does not need particles** at all. Radiation can travel through a vacuum.
- All objects take in and give out infrared radiation all the time. Hot objects radiate more infrared than cold objects. The amount of radiation given out or absorbed by an object depends on its temperature and on its surface.

Type of surface	As an emitter of radiation	As an absorber of radiation	Examples
Dull black	Good	Good	Cooling fans on the back of a refrigerator are dull black to radiate away more energy.
Bright shiny	Poor	Poor	Marathon runners, at the end of a race, wrap themselves in shiny blankets to prevent thermal transfer by radiation. Fuel storage tanks are sprayed with shiny silver paint to reflect radiation from the Sun.

Evaporation

- When particles break away from the surface of a liquid and form a vapour, the process is known as evaporation.

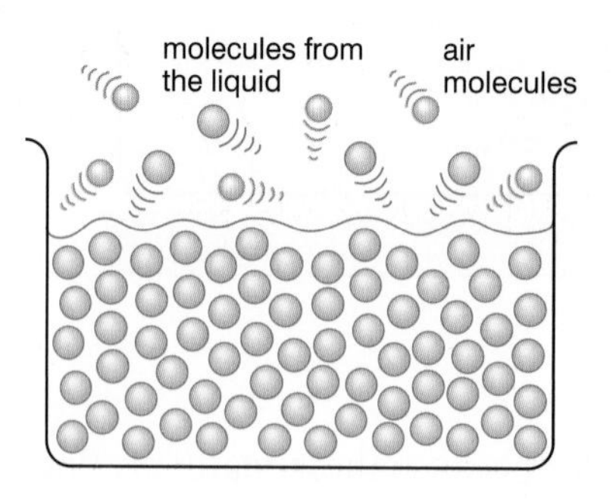

The more energetic molecules of the liquid escape from the surface. This reduces the average energy of the molecules remaining in the liquid and so the liquid cools down.

- Evaporation causes cooling. The evaporation of sweat helps to keep a body cool in hot weather. The cooling obtained in a refrigerator is also due to evaporation.

How can we keep our energy costs low?

- Heating a house can account for over 60% of a family's total energy bill. Reducing thermal energy transfer from the house to the outside can greatly reduce the amount of energy that is being used and save a lot of money.

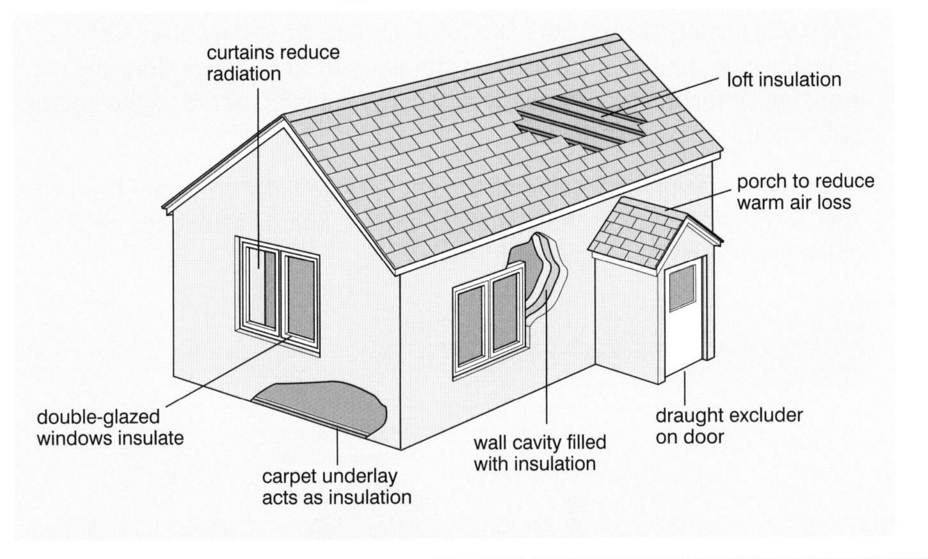

There are a number of ways of reducing wasteful energy transfer.

QUESTION SPOTTER

- Questions on insulation techniques frequently ask how the technique prevents energy transfer by conduction, convection or radiation.

Source of energy wastage	% of energy wasted	Insulation technique
Walls	35	*Cavity wall insulation.* Modern houses have cavity walls, that is, two single walls separated by an air cavity. The air reduces energy transfer by conduction but not by convection as the air is free to move within the cavity. Fibre insulation is inserted into the cavity to prevent the air from moving and so reduces convection.
Roof	25	*Loft insulation.* Fibre insulation is placed on top of the ceiling and between the wooden joists. Air is trapped between the fibres, reducing energy transfer by conduction and convection.
Floors	15	*Carpets.* Carpets and underlay prevent energy loss by conduction and convection. In some modern houses foam blocks are placed under the floors.
Draughts	15	*Draught excluders.* Cold air can get into the home through gaps between windows and doors and their frames. Draught excluder tape can be used to block these gaps.
Windows	10	*Double glazing.* Energy is transferred through glass by conduction and radiation. Double glazing has two panes of glass with a layer of air between the panes. It reduces energy transfer by conduction but not by radiation. Radiation can be reduced by drawing the curtains.

- One thing to consider when insulating a house is the balance of the cost of the insulation against the potential saving in energy costs. The **pay-back time** is the time it takes for the savings to repay the costs of installation. The different methods of insulation have very different pay-back times.
- The pay-back time is not the only thing to think about when considering different means of insulation. For instance, glazing reduces noise and condensation inside the home, and carpets provide increased comfort.

Means of insulation	Approximate pay-back time (years)
Cavity wall	5
Loft	2
Carpets	10
Draught excluders	1
Double glazing	20

IDEAS AND EVIDENCE

Reducing energy losses from a building has long-term advantages – less pollution and maintaining the stock of fossil fuels are just two. It can be easier to persuade people to conserve energy, however, by stressing the money-saving aspects – lower bills! How the data is presented is important to the overall impression.

Efficiency of energy transfers

- Energy transfers can be summarised using simple **energy transfer diagrams** or **Sankey diagrams**. The thickness of each arrow is drawn to scale to show the amount of energy.
- Energy is always conserved – the total amount of energy after the transfer must be the same as the total amount of energy before the transfer. Unfortunately, in nearly all energy transfers some of the energy will end up as 'useless' heat.
- In a power station only some of the energy originally produced from the fuel is transferred to useful electrical output. Energy **efficiency** can be calculated from the following formula:

$$\text{efficiency} = \frac{\text{useful energy output} \times 100\%}{\text{energy input}}$$

In a power station as much as 70% of the energy transfers do not produce useful energy. The power station is only 30% efficient.

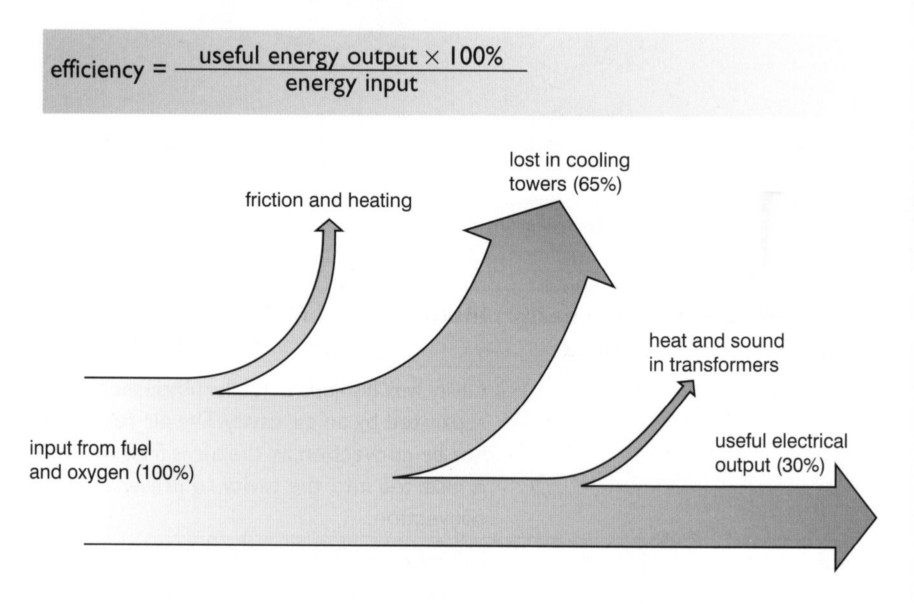

QUESTION SPOTTER

- You may be asked to label or complete a Sankey diagram like the one shown. Remember that the thickness of each section is directly related to its percentage contribution.

- Many power stations are now trying to make use of the large amounts of energy 'lost' in the cooling towers.

CHECK YOURSELF QUESTIONS

Q1 Why are several thin layers of clothing more likely to reduce thermal transfer than one thick layer of clothing?

Q2 Hot water in an open container transfers energy by evaporation. Explain how the loss of molecules from the surface of the liquid causes the liquid to cool.

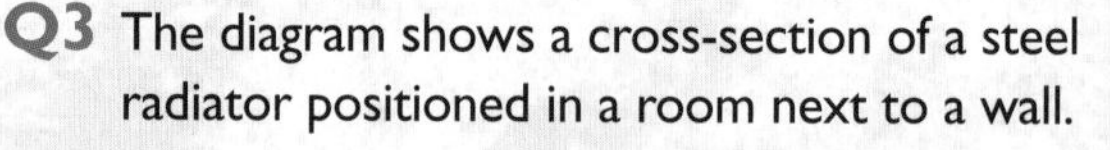

Q3 The diagram shows a cross-section of a steel radiator positioned in a room next to a wall.

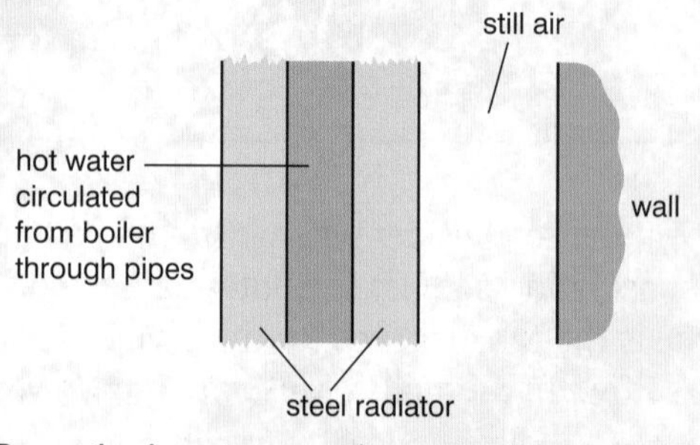

Describe how energy from the hot water reaches the wall behind the radiator.

Answers are on page 143.

REVISION SESSION 3

Specific heat capacity

How much energy is transferred during heating?

- When an object is heated, energy is transferred to it. As long as the object does not change state, for example melt or boil, this energy transfer will cause a **change in temperature**.
- The energy transferred during heating can be calculated using the formula:

> energy transferred = mass × specific heat capacity × temperature change
>
> $$E = m\,c\,\theta$$
>
> E = energy transferred in joules (J)
>
> m = mass in kilograms (kg)
>
> c = specific heat capacity in joules per kilogram per degree centigrade (J/kg/°C)
>
> θ = temperature change (°C)

- The formula shows that the energy transferred depends on:
  - the mass of the object (in kg). How much material is being heated?
  - the temperature change that happens (in °C). The bigger the temperature change, the more energy is transferred.
  - the specific heat capacity of the material (in J/kg/°C). This is a multiplying factor that depends on the material being heated.
- The **specific heat capacity**, c, is the energy required to raise the temperature of 1 kg of the material by 1 °C.

- You will be given the formula, but you might have to re-arrange it.

WORKED EXAMPLE

Calculate the energy transferred to 2 kg of water at 10 °C when it is boiled in a kettle. The specific heat capacity of water is 4200 J/kg/°C.

> The temperature change is from 10 °C to 100 °C
>
> Write down the formula: $E = m\,c\,\theta$
>
> Substitute the values: $E = 2 \times 4200 \times 90$
>
> Work out the answer and write down the unit: E = 756 000 J (= 756 kJ)

QUESTION SPOTTER

- Take care with units – if the mass is in g, the specific heat capacity should be J/g/°C,

So that's why the pie filling is so hot ...

- When a fruit pie, for example, is cooked, energy is transferred to the pastry and to the fruit filling at the same rate. However, the specific heat capacity of the filling is much lower than the specific heat capacity of the pastry. This means that, for the same amount of energy transferred to each, there will be a much greater temperature rise in the filling.

What about cooling down?

- When the energy is being **transferred away** from an object, that is it is **cooling down**, the formula is used in exactly the same way. This time, instead of calculating the energy transferred to the object, the answer refers to the energy transferred away from it.

> **A* EXTRA**
>
> - The properties of water are very strange. Not only does it require a great deal of heat to change its temperature, it also expands as it freezes. This makes ice less dense than liquid water, so ice floats (as the Titanic found out). This has been vital to evolution – life can survive at the bottom of ponds, where the water stays liquid, even when the surface has frozen.

Water is strange

- Water has a surprisingly high specific heat capacity. This means that a lot of energy has to be transferred to change the temperature of water significantly. This is important in several ways.
  - Water makes an excellent **coolant** for machines such as car engines. It can remove a lot of energy from the machine without boiling.
  - The temperature of the seas and oceans remains fairly steady, as huge energy transfers are needed to significantly change the temperature of that much water. This helps keep the planet at a fairly even temperature, which is good for living things.

> **IDEAS AND EVIDENCE**
>
> The evidence for, and possible consequences of, global warming cause a lot of discussion amongst scientists. Along with the more widely known carbon dioxide, water vapour is also a greenhouse gas. If global temperatures rise, more water vapour will evaporate from the oceans, putting more water vapour into the air. This may increase the greenhouse effect further. However, because this water vapour will form clouds, more sunlight may be reflected back into space, leading to a cooling effect.

CHECK YOURSELF QUESTIONS

Q1 Calculate the energy transferred when:

a 1.5 kg of copper is heated from 20 °C to 100 °C. The specific heat capacity of copper is 380 J/kg/°C.

b 1.5 kg of aluminium is heated from 20 °C to 100 °C. The specific heat capacity of aluminium is 880 J/kg/°C.

c Which material, copper or aluminium, would be best for making a saucepan? Explain your answer.

Q2 Explain why water is used to cool car engines.

Q3 A night storage heater contains a 50 kg block of concrete. At night, when electricity is cheaper, the concrete block is heated from 10 °C to 50 °C. During the day, the concrete transfers this stored energy to the room.

a How much energy is transferred to the concrete block during the night? The specific heat capacity of concrete is 800 J/kg/°C.

b How much energy is transferred to the room during the day?

c The specific heat capacity of copper is 380 J/kg/°C. Explain why copper would not be such a good material to store energy in the heater.

Answers are on page 143.

REVISION SESSION 4 Work, power and energy

Work

- **Work** is done when the application of a force results in movement. Work can only be done if the object or system has energy. When work is done energy is transferred.
- Work done can be calculated using the following formula:

> work done = force × distance moved
>
> $W = F s$
>
> W = work done in joules (J)
>
> F = force in newtons (N)
>
> s = distance moved in the direction of the force in metres (m)
>
> W
>
> F x s

In this position the gymnast is not doing any work against his body weight – he is not moving (he will be doing work pumping blood around his body though!).

The gymnast is doing work. He is moving upwards against the force of gravity. Energy is being transferred as he does the work.

WORKED EXAMPLES

1 A cyclist pedals along a flat road. She exerts a force of 60 N and travels 150 m. Calculate the work done by the cyclist.

Write down the formula:	$W = F s$
Substitute the values for F and s:	$W = 60 \times 150$
Work out the answer and write down the unit:	$W = 9000$ J

2 A person does 3000 J of work in pushing a supermarket trolley 50 m across a level car park. What force was the person exerting on the trolley?

Write down the formula with F as the subject:	$F = \frac{W}{s}$
Substitute the values for W and s:	$F = \frac{3000}{50}$
Work out the answer and write down the unit:	$F = 60$ N

Power

- **Power** is defined as the rate of doing work or the rate of transferring energy. The more powerful a machine is, the quicker it does a fixed amount of work or transfers a fixed amount of energy.
- Power can be calculated using the formula:

> $$\text{power} = \frac{\text{work done}}{\text{time taken}} = \frac{\text{energy transfer}}{\text{time taken}}$$
>
> $$P = \frac{W}{t}$$
>
> P = power in joules per second or watts (W)
>
> W = work done in joules (J)
>
> t = time taken in seconds (s)
>
> W
>
> P x t

- Calculations involving work and power are very common. In each case you will need to be able to change the subject of the equation, if necessary, and give the correct units.

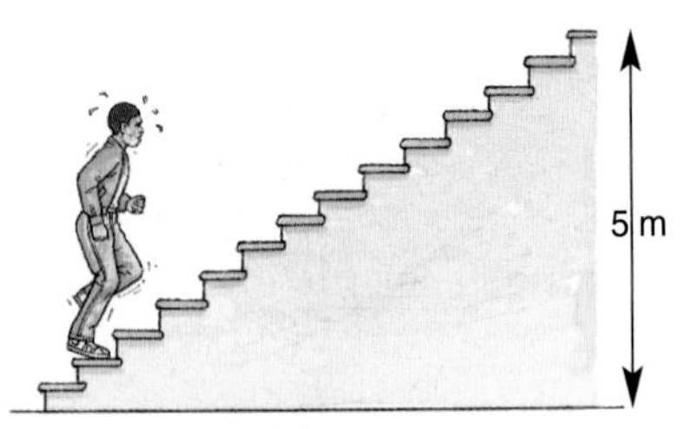

The student is lifting his body against the force of gravity, which acts in a vertical direction. The distance measured must be in the direction of the force (that is, the vertical height).

WORKED EXAMPLES

1 A crane does 20 000 J of work in 40 seconds. Calculate its power over this time.

Write down the formula:	$P = \frac{W}{t}$
Substitute the values for W and t:	$P = \frac{20\,000}{40}$
Work out the answer and write down the unit:	$P = 500\ \text{W}$

2 A student with a weight of 600 N runs up the flight of stairs shown in the diagram (left) in 4 seconds. Calculate the student's power.

Write down the formula for work done:	$W = F\,s$
Substitute the values for F and s:	$W = 600 \times 5 = 3000\ \text{J}$
Write down the formula for power:	$P = \frac{W}{t}$
Substitute the values for W and t:	$P = \frac{3000}{4} = 750\ \text{W}$

An object gains gravitational potential energy as it gains height. Work has to be done to increase the height of the object above the ground. Therefore:

gain in gravitational potential energy of an object = work done on that object against gravity.

Potential energy

- Stored, or hidden, energy is called **potential energy** (P.E.). If a spring is stretched, the spring will have potential energy. If a load is raised above the ground, it will have **gravitational potential energy**. If the spring is released or the load moves back to the ground, the stored potential energy is transferred to movement energy, which is called **kinetic energy** (K.E.).
- Gravitational potential energy can be calculated using the formula:

gravitational potential energy = mass × gravitational field strength × height

$$\text{P.E.} = m\,g\,h$$

P.E. = gravitational potential energy in joules (J)

m = mass in kilograms (kg)

g = gravitational field strength of 10 N/kg

h = height in metres (m)

The kinetic energy given to the stone when it is thrown is transferred to potential energy as it gains height and slows down. At the top of its flight practically all the kinetic energy will have been converted into gravitational potential energy. A small amount of energy will have been lost due to friction between the stone and the air.

WORKED EXAMPLE

A skier has a mass of 70 kg and travels up a ski lift a vertical height of 300 m. Calculate the change in the skier's gravitational potential energy.

Write down the formula:	P.E. = $m\,g\,h$
Substitute values for m, g and h:	P.E. = $70 \times 10 \times 300$
Work out the answer and write down the unit:	P.E. = 210 000 J or 210 kJ

Kinetic energy

- The kinetic energy of an object depends on its mass and its velocity. The kinetic energy can be calculated using the following formula:

$$\text{kinetic energy} = \frac{1}{2} \times \text{mass} \times \text{velocity}^2$$

$$\text{K.E.} = \frac{1}{2} mv^2$$

K.E. = kinetic energy in joules (J)

m = mass in kilograms (kg)

v = velocity in m/s

WORKED EXAMPLE

An ice skater has a mass of 50 kg and travels at a velocity of 5 m/s. Calculate the ice-skater's kinetic energy.

Write down the formula:	$\text{K.E.} = \frac{1}{2}mv^2$
Substitute the values for m and v:	$\text{K.E.} = \frac{1}{2} \times 50 \times 5 \times 5$
Work out the answer and write down the unit:	K.E. = 625 J

A* EXTRA

As a skier skis down a mountain the loss in potential energy should equal the gain in kinetic energy (assuming no other energy transfers take place, as a result of friction, for example). Calculations can then be performed using:

Loss in P.E. = Gain in K.E.
($mgh = \frac{1}{2}mv^2$)

CHECK YOURSELF QUESTIONS

Q1 50 000 J of work are done as a crane lifts a load of 400 kg. How far did the crane lift the load? (Gravitation field strength, g, is 10 N/kg.)

Q2 A student is carrying out a personal fitness test.

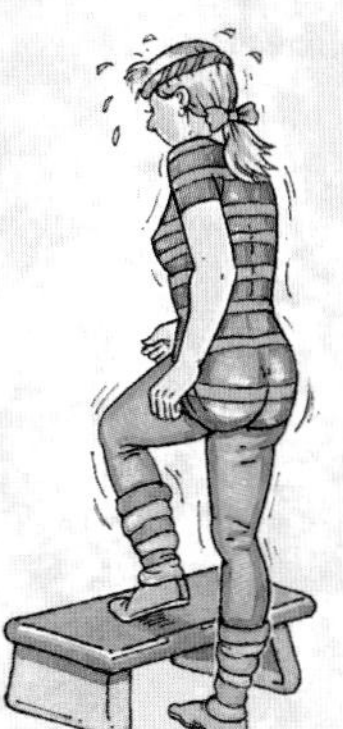
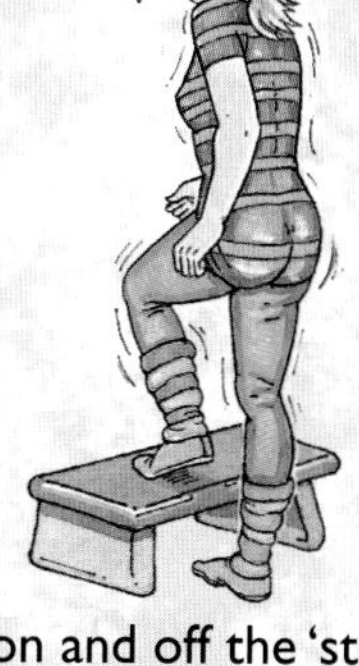

She steps on and off the 'step' 200 times. She transfers 30 J of energy each time she steps up.

a Calculate the energy transferred during the test.

b She takes 3 minutes to do the test. Calculate her average power.

Q3 A child of mass 35 kg climbed a 30 m high snow-covered hill.

a Calculate the change in the child's potential gravitational energy.

b The child then climbed onto a lightweight sledge and slid down the hill. Calculate the child's maximum speed at the bottom of the hill. (Ignore the mass of the sledge.)

c Explain why the actual speed at the bottom of the hill is likely to be less than the value calculated in part **b**.

Answers are on page 144.

UNIT 6: DESCRIBING WAVES

REVISION SESSION 1

The properties of waves

There are two types of waves

- **Longitudinal waves.** This type of wave can be shown by pushing and pulling a spring. The spring stretches in places and squashes in others. The stretching produces regions of **rarefaction**, whilst the squashing produces regions of **compression**. Sound is an example of a longitudinal wave.
- **Transverse waves.** In a transverse wave the vibrations are at right angles to the direction of motion. Light, radio and other electromagnetic waves are transverse waves. Water waves are often used to demonstrate the properties of waves because the **wavefront** of a water wave is easy to see. A wavefront is the moving line that joins all the points on the crest of a wave.

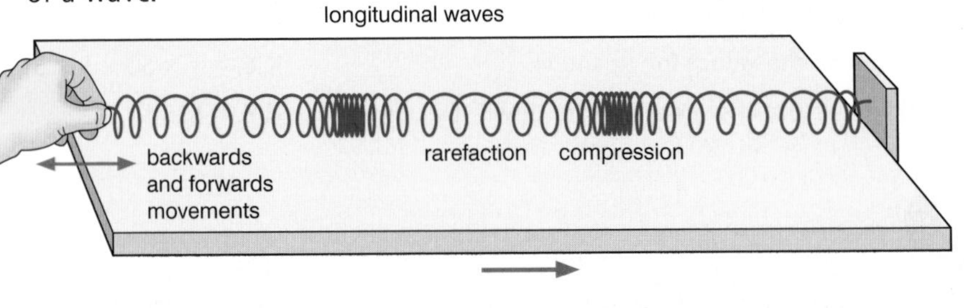

Longitudinal and transverse waves are made by vibrations.

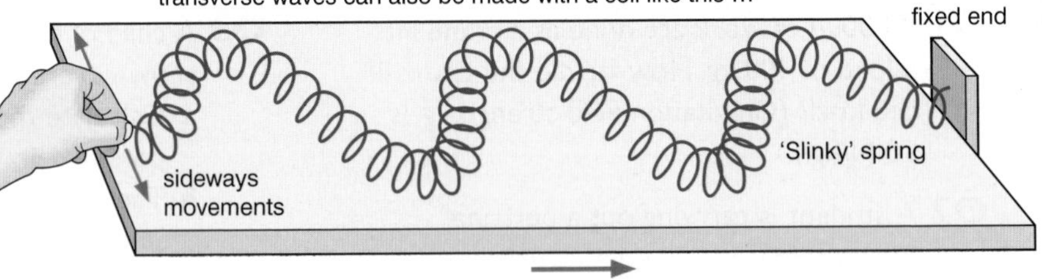

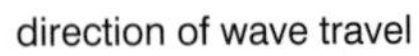

- You will often be asked to label a transverse wave showing features such as crest, trough, amplitude and wavelength.

What features do all waves have?

- Waves have a **repeating shape** or pattern.
- Waves **transfer energy** without moving material along.
- Waves have a wavelength, frequency and amplitude.
  - **Wavelength** – the length of the repeating pattern.
  - **Frequency** – the number of repeated patterns that go past any point each second.
  - **Amplitude** – the maximum displacement of the medium's vibration. In transverse waves, this is half the crest-to-trough height.

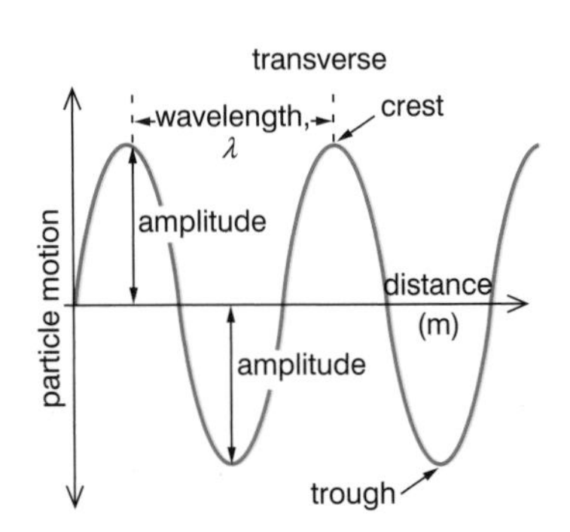

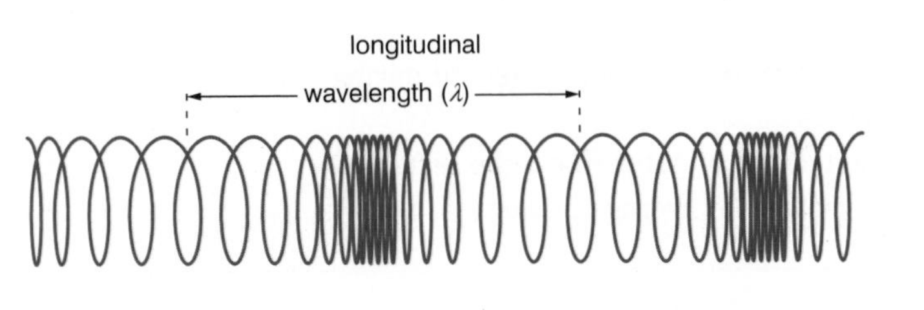

- The speed a wave travels at depends on the substance or **medium** it is passing through.
- **The speed of a wave in a given medium is constant.** If you change the wavelength the frequency *must* change as well. Speed, frequency and wavelength of a wave are related by the equation:

WORKED EXAMPLES

wave speed = frequency × wavelength

$v = f \times \lambda$

v = wave speed, usually measured in metres/second (m/s)

f = frequency, measured in cycles per second or hertz (Hz)

λ = wavelength, usually measured in metres (m)

- Calculations involving the wave equation are very common. You will need to be able to remember the equation and change its subject if required.

1 A loudspeaker makes sound waves with a frequency of 300 Hz. The waves have a wavelength of 1.13 m. Calculate the speed of the sound waves.

Write down the formula:	$v = f \times \lambda$
Substitute the values for f and λ:	$v = 300 \times 1.13$
Work out the answer and write down the unit:	$v = 339$ m/s

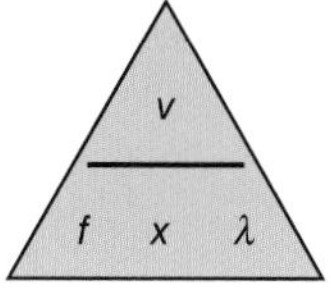

2 A radio station broadcasts on a wavelength of 250 m. The speed of the radio waves is 3×10^8 m/s. Calculate the frequency.

Write down the formula with f as the subject:	$f = \frac{v}{\lambda}$
Substitute the values for v and λ:	$f = \frac{3 \times 10^8}{250}$
Work out the answer and write down the unit:	$f = 1\,200\,000$ Hz or 1200 kHz

Reflection, refraction and diffraction

- When a wave hits a barrier the wave will be **reflected**. If it hits the barrier at an angle then the **angle of reflection** will be **equal** to the **angle of incidence**. **Echoes** are a common consequence of the reflection of sound waves.

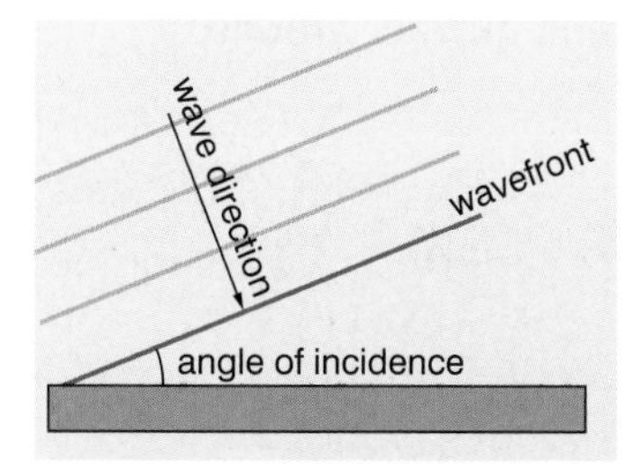

Waves hit a barrier. The angle between a wavefront and the barrier is the angle of incidence.

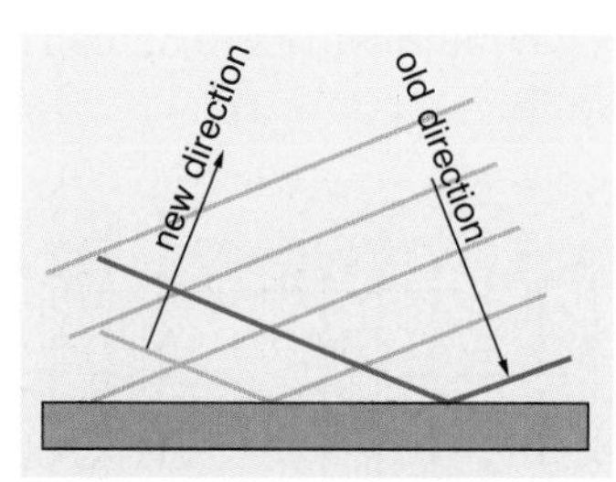

The waves bounce off.

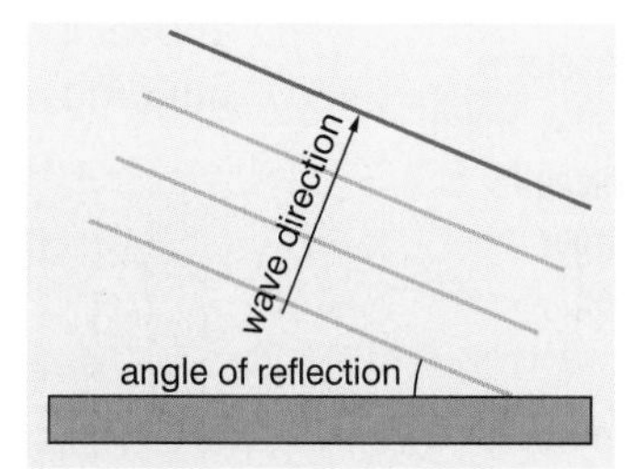

The angle of reflection is the same as the angle of incidence.

- When a wave moves from one medium into another, it will either speed up or slow down. This is known as **refraction**. When a wave **slows down**, the wavefronts crowd together – the **wavelength gets smaller**. When a wave **speeds up**, the wavefronts spread out – the **wavelength gets larger**.

When waves slow down, their wavelength gets shorter.

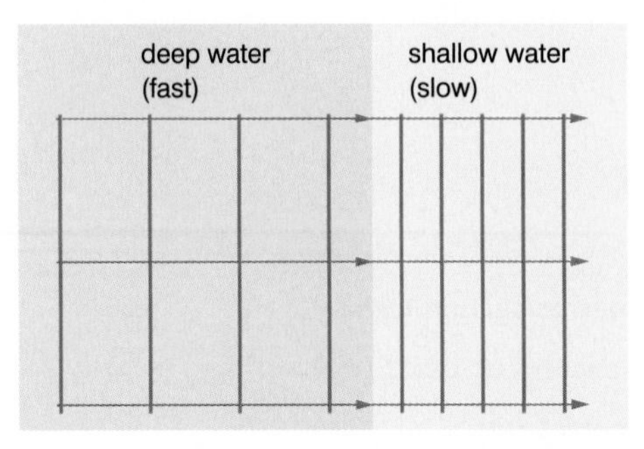

- If a wave enters a new medium at an angle then the wavefronts also change direction. Refraction happens whenever there is a change in wave speed. Water waves are slower in shallower water than in deep water, so water waves will refract when the depth changes.

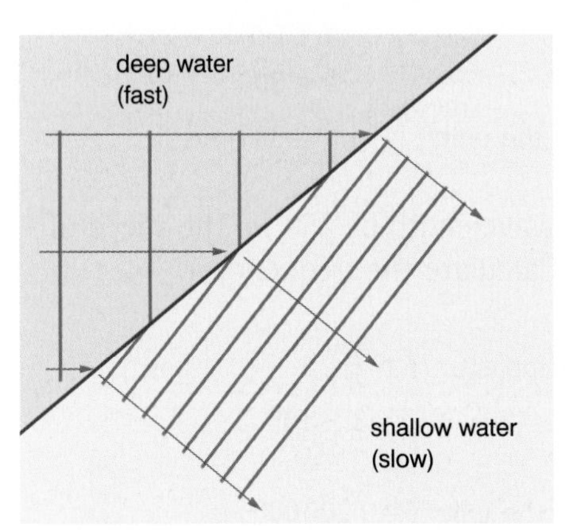

If waves cross into a new medium at an angle, their wavelength and direction changes.

A* EXTRA

- High notes from a CD track have wavelengths of around 10–20 cm. For these to diffract efficiently and spread out as they leave the opening of the loudspeaker, a speaker with a similar diameter is used. For lower notes, which have larger wavelengths, a speaker with a larger diameter is needed.

- Wavefronts change shape when they pass the edge of an obstacle or go through a gap. This process is known as **diffraction.** Diffraction is strong when the width of the gap is similar in size to the wavelength of the waves.
- Diffraction is a problem in communications when radio and television signals are transmitted through the air in a narrow beam. Diffraction of the wavefront means that not all the energy transmitted with the wavefront reaches the receiving dishes. On the other hand, diffraction allows long-wave radio waves to spread out and diffract around buildings and hills.

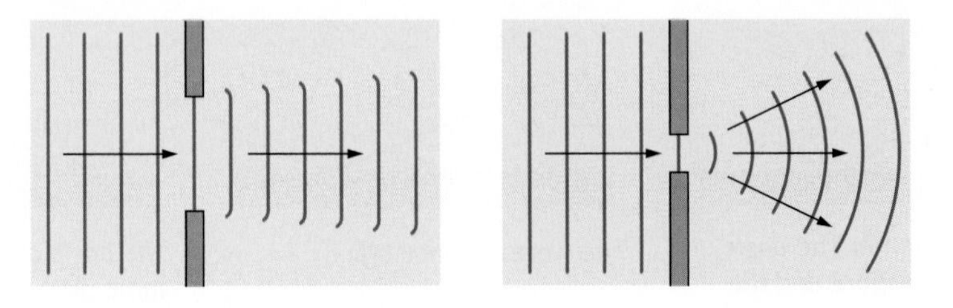

Diffraction is most noticeable when the size of the gap equals the wavelength of the waves.

Seismic waves

- Earthquakes make waves that travel right through the Earth. These waves are called **seismic waves**. There are longitudinal waves called **P-waves** (primary) and transverse waves called **S-waves** (secondary).
- The waves travel through rock and are partially reflected at the boundaries between different types of rock. Instruments called **seismometers** detect the waves. Monitoring these seismic waves after earthquakes has provided geologists with evidence that the Earth is made up of layers and that part of the core is liquid.

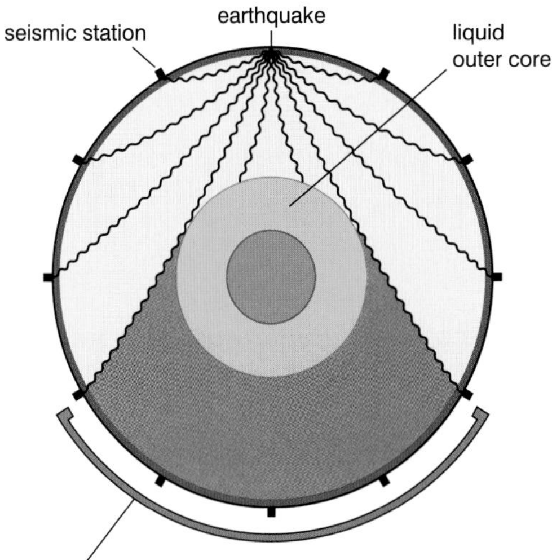

S-waves can travel through solid rock but not through liquid. P-waves can travel through solid and liquid rock. Only the P-waves from an earthquake, not the S-waves, are received at seismic stations on the opposite side of the world.

Check Yourself Questions

Q1 The diagram shows a trace of a sound wave obtained on a cathode ray oscilloscope.

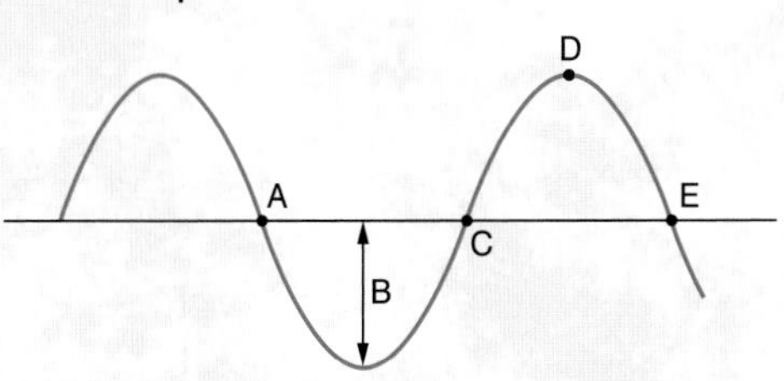

a Which letter shows the crest of the wave?

b The wavelength is the distance between which two letters?

c Which letter shows the amplitude?

d The frequency of the wave is 512 Hz. How many waves are produced each second?

Q2 Radio waves of frequency 900 MHz are used to send information to and from a portable phone. The speed of the waves is 3×10^8 m/s. Calculate the wavelength of the waves. (1 MHz = 1 000 000 Hz, 3×10^8 = 300 000 000.)

Q3 The speed of sound is approximately 340 m/s. Calculate the wavelength of middle C, which has a frequency of 256 Hz.

Answers are on page 144.

The electromagnetic spectrum

Electromagnetic waves

A prism splits white light into the colourful spectrum of visible light.

- The **electromagnetic spectrum** is a 'family' of waves. Electromagnetic waves all travel at the same speed in a vacuum, i.e. the speed of light, 300 000 000 m/s.
- **White light** is a mixture of different colours and can be split by a prism into the **visible spectrum.** All the different colours of light travel at the same speed in a vacuum but they have different frequencies and wavelengths. When they enter glass or perspex they all slow down, but by different amounts. The different colours are therefore refracted through different angles. Violet is refracted the most, red the least.
- The visible spectrum is only a small part of the full electromagnetic spectrum.

The complete electromagnetic spectrum.

Type of wave	gamma rays	X rays	ultraviolet	visible	infrared	microwaves	TV and radio waves
frequency	high						low
wavelength	low						high
use	killing cancer cells	to look at bones	sun tan beds	photography	TV remote controls	cooking	transmission of TV and radio

- Questions often require you to remember which waves in the electromagnetic spectrum have the greatest frequency or greatest wavelength.

A* EXTRA

- The energy associated with an electromagnetic wave depends on its frequency. The waves with the higher frequencies are potentially the more hazardous.

The waves you cannot see

- **Gamma rays** are produced by radioactive nuclei. They transfer more energy than X-rays. Gamma rays are frequently used in radiotherapy to kill cancer cells. Radioactive substances that emit gamma rays are used as tracers (see Unit 9, Revision Session 4).
- **X-rays** are produced when high-energy electrons are fired at a metal target. Bones absorb more X-rays than other body tissue. If a person is placed between the X-ray source and a photographic plate, the bones appear to be white on the developed photographic plate compared with the rest of the body. X-rays have very high energy and can damage or destroy body cells. They may also cause cancer. Cancer cells absorb X-rays more readily than normal healthy cells and so X-rays are also used to treat cancer.
- **Ultraviolet radiation** (UV) is the component of the Sun's rays that gives you a suntan. UV is also created in fluorescent light tubes by exciting the atoms in a mercury vapour. The UV radiation is then absorbed by the coating on the inside of the fluorescent tube and re-emitted as visible light. Fluorescent tubes are more efficient than light bulbs because they do not depend on heating and so more energy is available to produce light.

- All objects give out **infrared radiation** (IR). The hotter the object is, the more radiation it gives out. Thermograms are photographs taken to show the infrared radiation given out from objects. Infrared radiation grills and cooks our food in an ordinary oven and is used in remote controls to operate televisions and videos.
- **Microwaves** are high-frequency radio waves. They are used in radar to find the position of aeroplanes and ships. Metal objects reflect the microwaves back to the transmitter, enabling the distance between the object and the transmitter to be calculated. Microwaves are also used for cooking. Water particles in food absorb the energy carried by microwaves. They vibrate more and get much hotter. Microwaves penetrate several centimetres into the food and so speed up the cooking process.
- **Radio waves** have the longest wavelengths and lowest frequencies. **UHF** (ultra-high frequency) waves are used to transmit television programmes to homes. **VHF** (very high frequency) waves are used to transmit local radio programmes. **Medium** and **long** radio waves are used to transmit over longer distances because their wavelengths allow them to diffract around obstacles such as buildings and hills. Communication satellites above the Earth receive signals carried by high-frequency (**short-wave**) radio waves. These signals are amplified and re-transmitted to other parts of the world.

A* EXTRA

- Infrared radiation is absorbed by the surface of the food and the energy is spread through the rest of the food by conduction.
- In contrast, microwaves penetrate a few centimetres into food and then the energy is transferred throughout the food by conduction.

CHECK YOURSELF QUESTIONS

Q1 This is a list of types of wave:

gamma *infrared* *light* *microwaves*
radio *ultraviolet* *X-rays*

Choose from the list the type of wave that best fits each of these descriptions.

a Stimulates the sensitive cells at the back of the eye.
b Necessary for a suntan.
c Used for *rapid* cooking in an oven.
d Used to take a photograph of the bones in a broken arm.
e Emitted by a video remote control unit.

Q2 Gamma rays are part of the electromagnetic spectrum. Gamma rays are useful to us but can also be very dangerous.

a Explain how the properties of gamma rays make them useful to us.
b Explain why gamma rays can cause damage to people.
c Give one difference between microwaves and gamma rays.
d Microwaves travel at 300 000 000 m/s. What speed do gamma rays travel at?

Answers are on page 144.

Light reflection and refraction

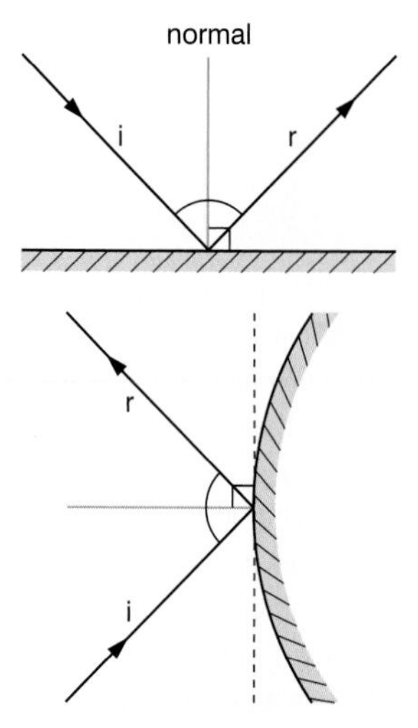

The angles of incidence and reflection are the same when a mirror reflects light. This type of curved mirror is known as a convex mirror.

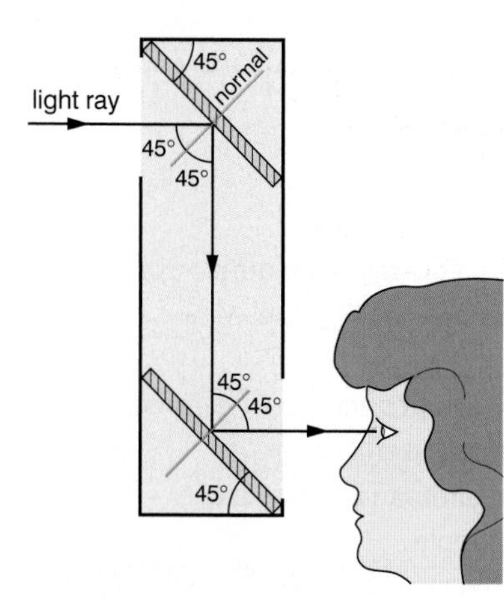

A periscope uses reflection to allow you to see above your normal line of vision – or even round corners.

Light reflects

- Light rays are reflected from mirrors in such a way that

  angle of incidence (*i*) = angle of reflection (*r*)

  The angles are measured to an imaginary line at 90° to the surface of the mirror. This line is called the **normal**. With a curved mirror it is difficult to measure the angle between the ray and the mirror.

- When you look in a plane mirror you see an **image** of yourself. The image is said to be **laterally inverted** because if you raise your right hand your image raises its left hand. The image is formed as far behind the mirror as you are in front of it and is the same size as you. The image cannot be projected onto a screen. It is known as a **virtual image**.

Uses of mirrors

- In a plane mirror the image is always the same size as the object. Examples of plane mirrors include household 'dressing' mirrors, dental mirrors for examining teeth, security mirrors for checking under vehicles, periscope.
- Close to a **concave mirror** the image is **larger** than the object – these mirrors **magnify**. They are used in torches and car headlamps to produce a beam of light, make-up and shaving mirrors, satellite dishes.
- The image in a **convex mirror** is always **smaller** than the object Examples are a car driving mirror, shop security mirror.

Light refracts

- Light waves **slow down** when they travel from air into glass. If they are at an angle to the glass, they bend towards the normal. When the light rays travel out of the glass into the air, their speed increases and they bend away from the normal.

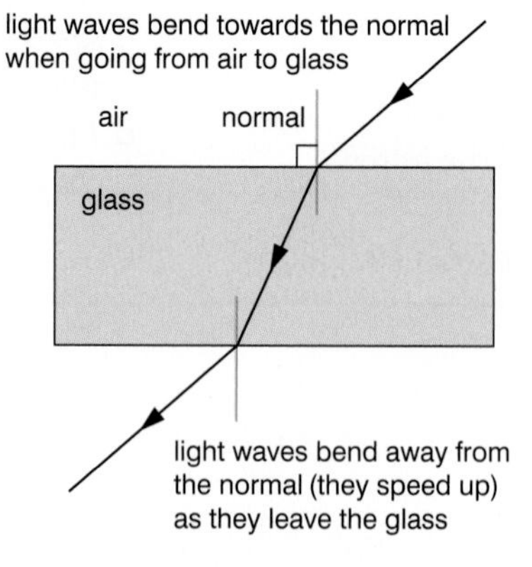

- The **refractive index** of a material indicates how strongly the material changes the speed of the light. It is calculated as:

$$\text{refractive index} = \frac{\text{speed of light in a vacuum}}{\text{speed of light in the medium}}$$

- Refraction is observed when a triangular prism (a glass or plastic block) is used to obtain the visible spectrum (see page 78). Lenses also refract light rays.

Colours

- The **primary** colours for mixing light are **red, green** and **blue**. Mixing **two** of these colours together produces a **secondary** colour. The secondary colours are **magenta** (mixed from red and blue), **cyan** (mixed from green and blue) and **yellow** (mixed from red and green). Mixing **all three** primary colours produces **white** light.
- A coloured filter allows only light of its own colour to pass through (or the primary colours that make up that colour).

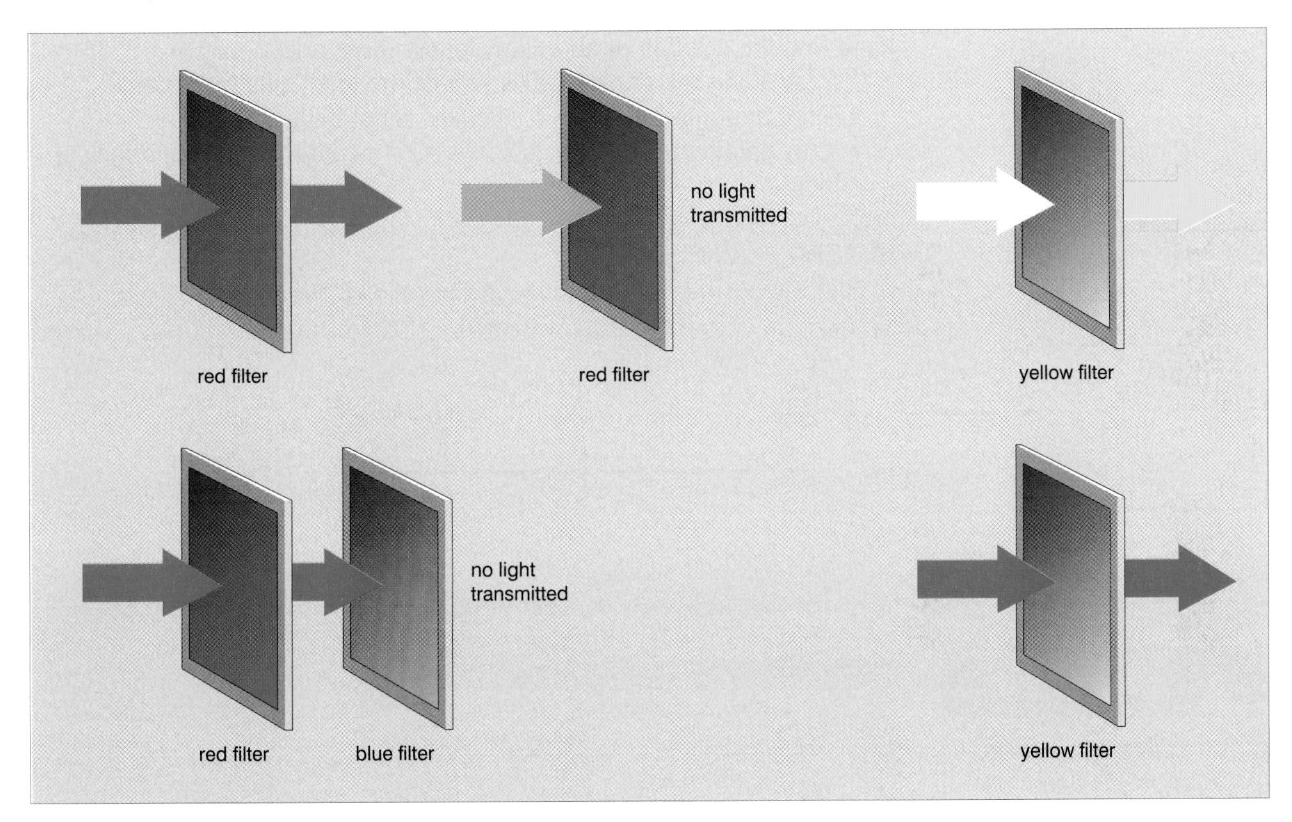

- An object appears to be a particular colour because it reflects *only* that colour (or the primary colours that make up that colour). Any other light is **absorbed**. If all the light is absorbed, then an object appears **black**.

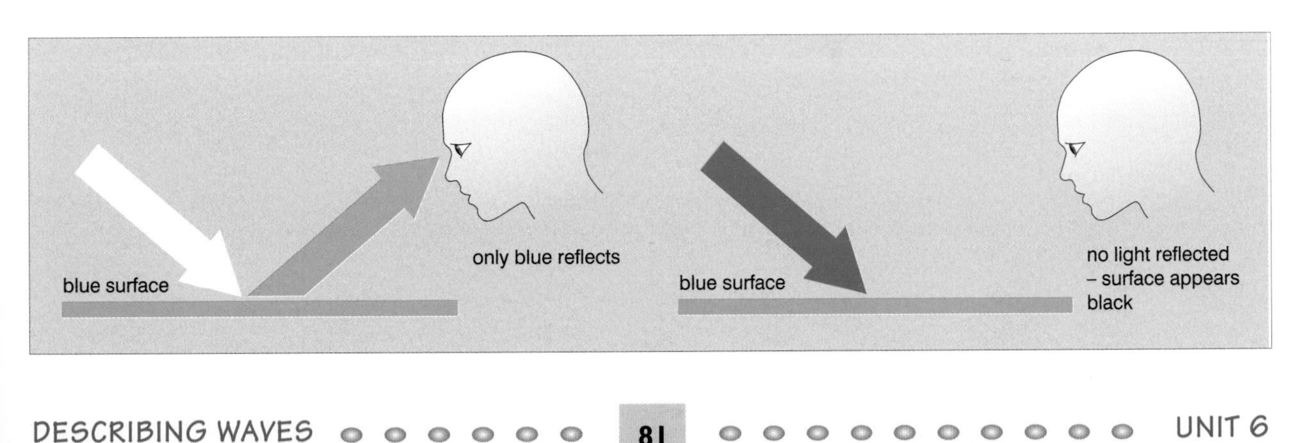

Lenses

- **Convex (converging, positive) lenses** cause parallel rays of light to **converge** to the principal focus. They are used to form **images** in magnifying glasses, cameras, telescopes, binoculars, microscopes, film projectors and spectacles.

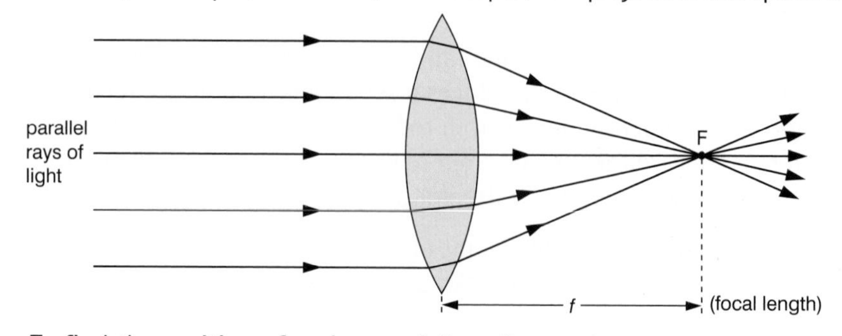

- To find the position of an image, follow these rules:
  - A ray along the principal axis (the centre line) is not deflected.
  - A ray through the centre of the lens is not deflected.
  - A ray parallel to the principal axis is refracted through the principal focus.

WORKED EXAMPLE

Find the position of the image in the following diagram.

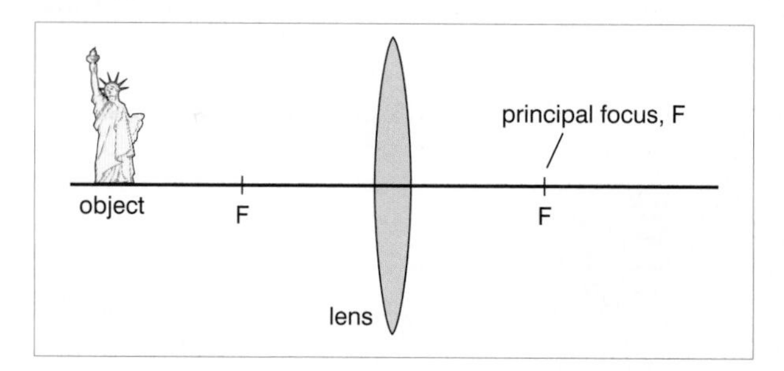

A* EXTRA

- If the diagram is drawn to scale, the size of the image can also be found.

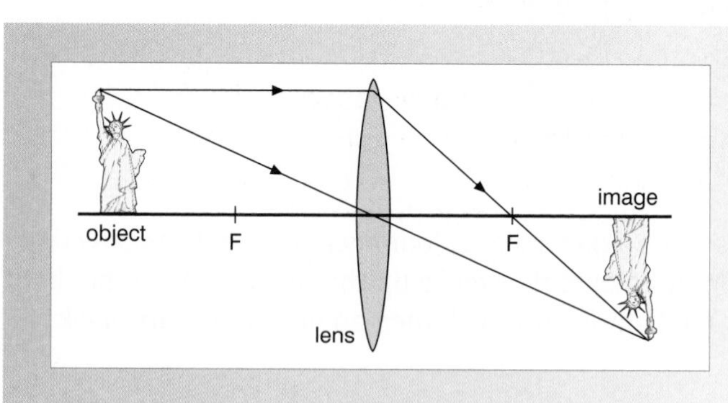

- **Concave (diverging, negative) lenses** make parallel rays of light **diverge** as if they were coming from the principal focus. You see concave lenses in spectacles, telescopes and correcting lenses in all optical equipment.

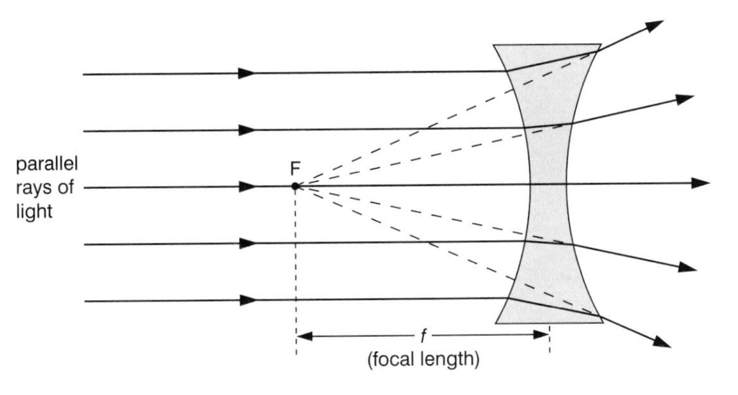

Total internal reflection

- When rays of light pass from a **slow medium** to a **faster medium** they move **away** from the normal.
- As the angle of incidence increases, an angle is reached at which the light rays will have to leave with an angle of refraction greater than 90°! These rays cannot refract, so they are entirely reflected back inside the medium. This process is known as **total internal reflection.**
- The angle of incidence at which all refraction stops is known as the **critical angle** for the material. The critical angle of glass is 42°, the critical angle of water is 49°.
- Total internal reflection is used in **fibre-optic cables.** These are made up of large numbers of very thin, glass fibres. The light continues along the fibres by being constantly internally reflected. Fibre-optic cables are used in medical endoscopes for internal examination of the body.
- Telephone and TV communications systems are increasingly relying on fibre optics instead of the more traditional copper cables. Fibre-optic cables do not use electricity and the signals are carried by infrared rays. The signals are very clear as they do not suffer from electrical interference. Other advantages are that they are cheaper than the copper cables and can carry thousands of different signals down the same fibre at the same time (see Unit 7).

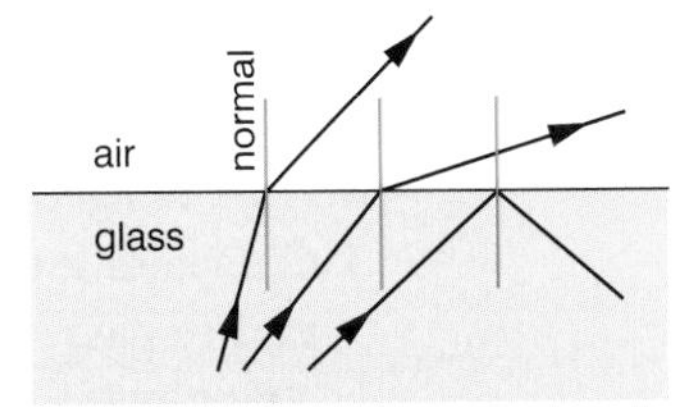

Total internal reflection occurs when a ray of light tries to leave the glass. If the angle of incidence equals or is greater than the critical angle the ray will be totally internally reflected.

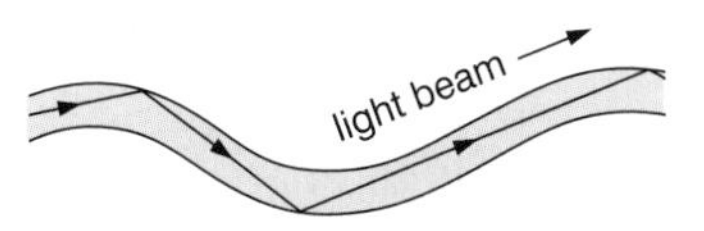

Light does not escape from the fibre because it always hits it at an angle greater than the critical angle and is internally reflected.

IDEAS AND EVIDENCE

Optical fibres can transfer more information more quickly than copper cables. For example, they are used for broadband internet access. However, not everyone has access to broadband connections, often due to their location. For instance, it is easier and more economic to replace copper cables with optical fibres in built-up areas than in rural districts.

QUESTION SPOTTER

- Questions on total internal reflection are common. Typically you will have to draw or interpret simple ray diagrams in objects such as binoculars, periscopes, bicycle reflectors or endoscopes.

CHECK YOURSELF QUESTIONS

Q1 **a** Rays of light can be reflected and refracted. State one difference between reflection and refraction.

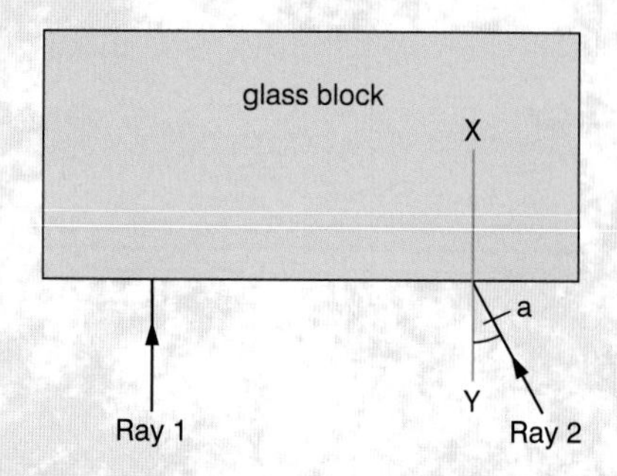

b The diagram shows a glass block and two rays of light.

i Complete the paths of the two rays as they pass into and then out of the glass block.

ii What name is given to the angle marked a?

iii What name is given to the line marked XY?

Q2 **a** Tom looks into the mirror. E is his eye and X an object.

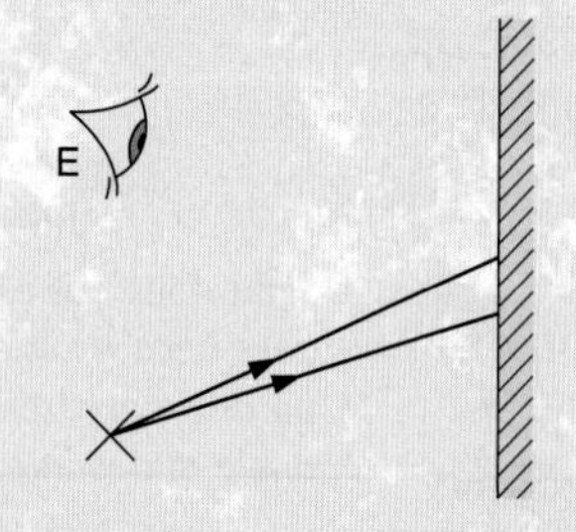

Complete the ray diagram to show clearly how Tom sees an image of X.

b Jane uses two mirrors to look at the back of her head. Explain how this works.

Q3 The diagram shows light entering a prism. Total internal reflection takes place at the inner surfaces of the prism.

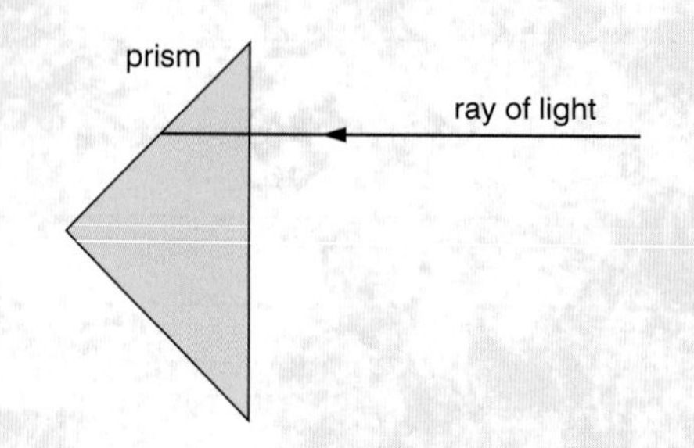

a Complete the path of the ray.

b Suggest one use for a prism like this.

c Complete the table about total internal reflection. Use T for total internal reflection or R for refraction.

angle of incidence/ degrees	total internal reflection (T) or refraction (R)
36	
42 (critical angle)	
46	

Answers are on page 145.

Sound waves, resonance and musical instruments

Properties of sound waves

- Sound waves travel at about 340 m/s in the air – much slower than the speed of light. This explains why you almost always see the flash of lightning before hearing the crash of the thunder.
- Sound is caused by vibrations, and travels as **longitudinal waves.** The compressions and rarefactions of sound waves result in small differences in air pressure.
- Sound waves travel faster through liquids than through air. Sound travels fastest through solids. This is because particles are linked most strongly in solids.
- High-**pitch** sounds have a high frequency whereas low-pitch sounds have a low frequency. The human ear can detect sounds with pitches ranging from 20 Hz to 20 000 Hz. Sound with frequencies above this range is known as **ultrasound**.
- Loud sounds have high amplitude whereas quiet sounds have low amplitude. The loudness of sounds can be compared using **decibels.**

WORKED EXAMPLE

The diagram shows a displacement–time graph for a sound wave.
Calculate: a the amplitude of the sound
b the frequency of the sound.

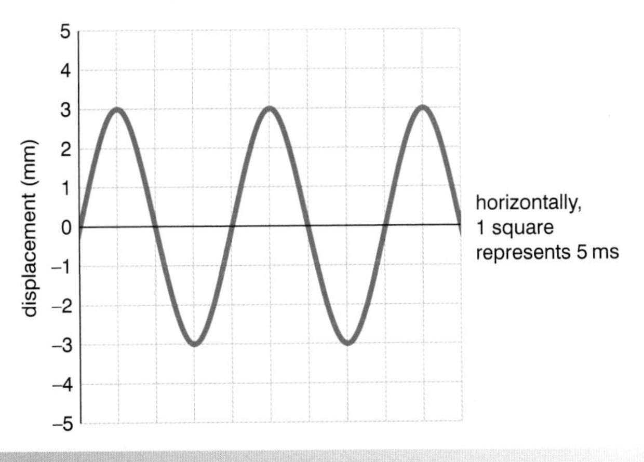

a The amplitude is the maximum displacement from the mean position, so can be read straight from the graph.

Amplitude = 3 mm.

b From the graph, work out the time taken for one cycle:

One cycle covers 4 squares and each square represents 5 ms, so the time for one cycle is 20 ms.

Convert to seconds: 1 ms = 0.001 seconds, so 20 ms = 0.020 s

The frequency is the number of cycles per second $= \frac{1}{0.020} = 50$ Hz.

QUESTION SPOTTER

- 1 second = 1000 ms. Take care not to confuse ms (milliseconds) with m/s (metres per second).

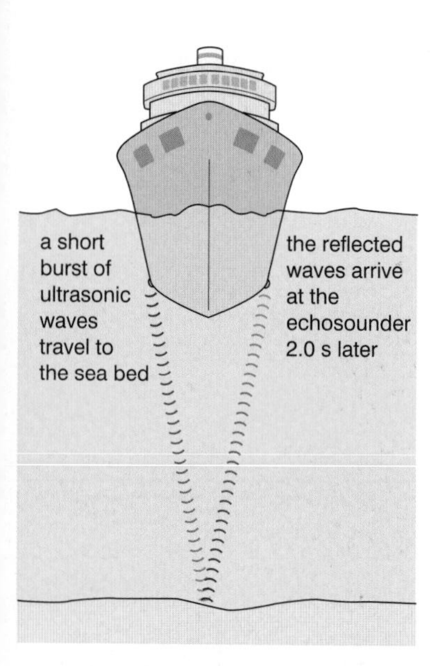

How do we use ultrasound?

- Ultrasound is used in echo sounding or SONAR (SOund NAvigation and Ranging). An echo sounder on a ship sends out ultrasonic waves which reflect off the bottom of the sea. The depth of the water can be calculated from the time taken for the echo to be received and the speed of sound in water.

WORKED EXAMPLE

A ship sends out an ultrasound wave and receives an echo in 2 seconds. If the speed of sound in water is 1500 m/s, how deep is the water?

Write down the formula:	speed = distance/time or $v = s/t$
Rearrange to make s the subject:	$s = v \times t$
Substitute the values for v and t:	$s = 1500 \times 2 = 3000$ m

This is the distance travelled by the sound wave.
Therefore the depth of the water must be half this.
Depth = 1500 m.

A* EXTRA

- Ultrasound waves are reflected from the different layers of tissues in the body and so can produce quite clear images. They have lower energy than X-rays and so are less hazardous.
- However, they can be used to break down solid material such as kidney stones when they are highly focused.

- Ultrasound is used in medicine. Dense material, such as bone, reflects more ultrasound waves than less dense material, such as skin and tissue. Ultrasound waves have lower energy than X-rays and so are less likely to damage healthy cells.

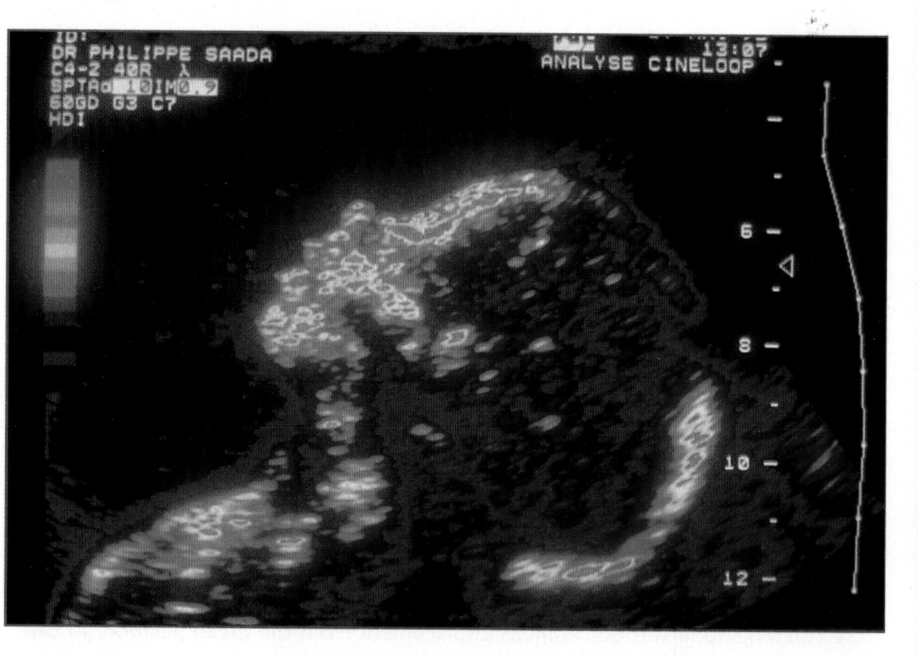

The foetus reflects the ultrasound waves more than the surrounding amniotic fluids.

What is resonance?

- All objects vibrate with a characteristic **natural frequency**. When an object is **forced** to vibrate at its natural frequency, the amplitude of its vibration can grow very large. This effect is called **resonance**. For example, a pendulum will make much larger swings if it given an extra push timed to match the natural frequency of the pendulum.

- Resonance can have *undesirable* effects. For example, the seismic waves from an earthquake force the buildings at the surface to vibrate. If the frequency of the vibrations is at the natural frequency of the building, then the amplitude of the vibration increases, more energy is transferred to the building and the building will probably fall down.
- Resonance can also be *useful*. In a microwave ovens, the microwaves cause water molecules in the food to vibrate more strongly and this heats the food (see Revision Session 2).
- In many musical instruments, resonance is used to make the sound much louder at a particular frequency.

Musical instruments

- The natural frequency of a **vibrating string** (for example, on a guitar, violin, harp or piano) depends on:
  - The length of the string
  - The tension in the string
  - The mass of the string.

  Higher notes are produced by strings that are shorter, tighter and lighter.
- When a string vibrates, a stationary wave pattern of **nodes** and **antinodes** is formed.
- At a node, the amplitude of the vibration is zero.
  At an antinode, the amplitude is a maximum.

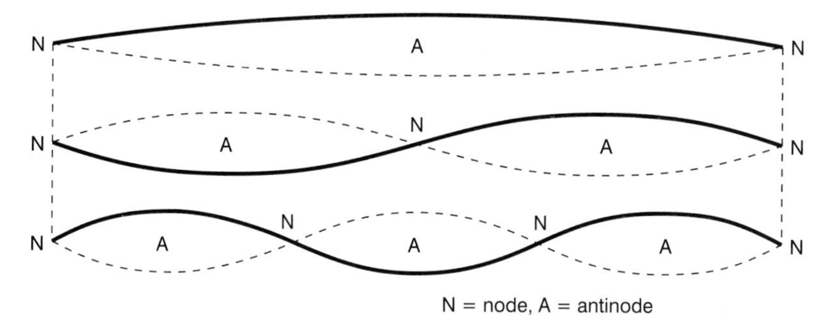

N = node, A = antinode

- In a vibrating **air column**, such as a flute or an organ pipe, the natural frequency decreases if the column is lengthened, so higher notes come from shorter pipes.

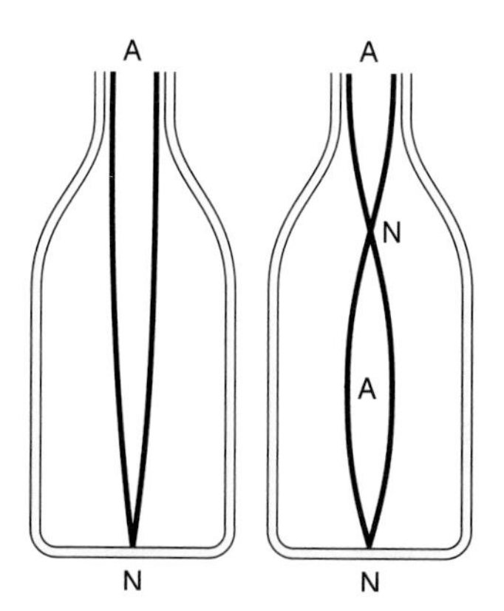

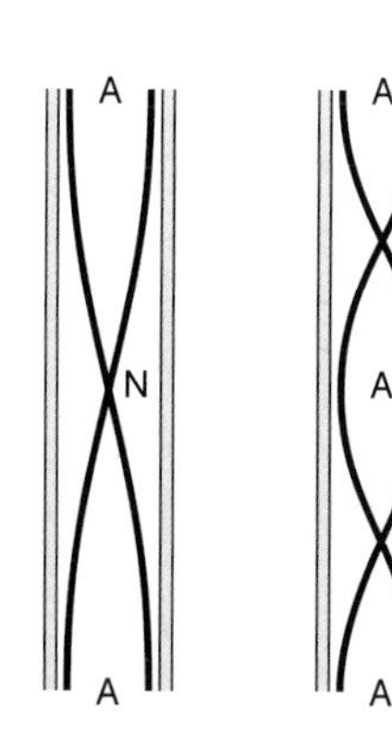

- When a string or air column vibrates, the different patterns of nodes and antinodes all occur together. The *relative strength* of these patterns decides the *quality*, or 'sound' of the note. This is how you can tell the difference between a guitar and a harp playing the same note.

CHECK YOURSELF QUESTIONS

Q1 **a** **i** What causes a sound?
ii Explain how sound travels through the air.
b Astronauts in space cannot talk directly to each other. They have to speak to each other by radio. Explain why this is so.
c Explain why sound travels faster through water than through air.
d It is often said that opera singers can sing a high note which will shatter a wine glass. Explain how this might happen.

Q2 Ultrasonic waves are longitudinal waves.
a What does the word 'ultrasonic' mean?
b What does the word 'longitudinal' mean?
c The waves travel through carbon dioxide more slowly than through air. How do the frequency and wavelength change when ultrasonic waves pass from air to carbon dioxide?

Q3 A fishing boat was using echo sounding to detect a shoal of fish. Short pulses of ultrasound were sent out from the boat. The echo from the shoal was detected 0.5 seconds later. How far away from the boat was the shoal of fish? (Sound waves travel through water at a speed of 1500 m/s.)

Answers are on page 146.

REVISION SESSION 5

Interference

What is interference?

- Interference effects happen when two (or more) sets of waves overlap. Where the sets of waves reinforce each other the overall amplitude increases – this is often called **constructive interference**. Where the sets of waves cancel out the overall amplitude is reduced – often called **destructive interference**.

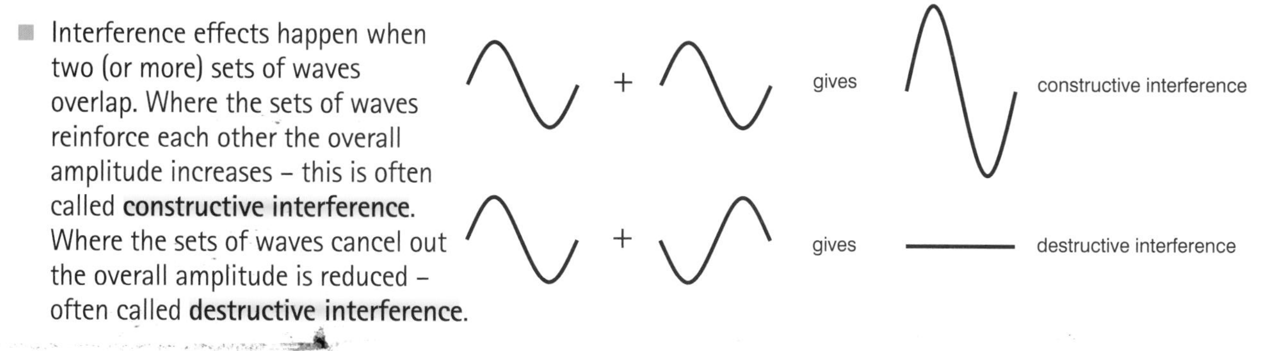

How do we demonstrate interference in the laboratory?

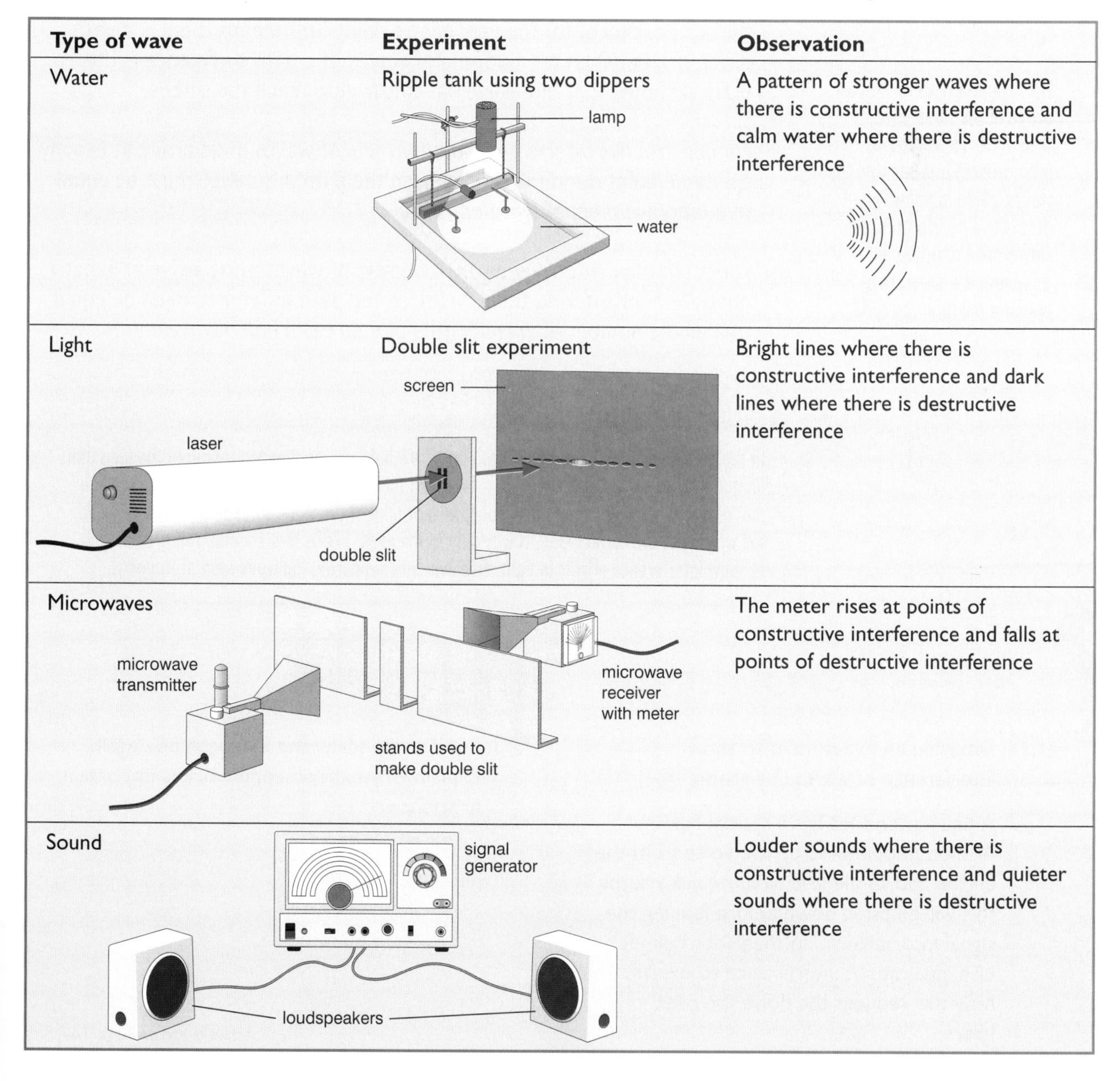

Type of wave	Experiment	Observation
Water	Ripple tank using two dippers	A pattern of stronger waves where there is constructive interference and calm water where there is destructive interference
Light	Double slit experiment	Bright lines where there is constructive interference and dark lines where there is destructive interference
Microwaves		The meter rises at points of constructive interference and falls at points of destructive interference
Sound		Louder sounds where there is constructive interference and quieter sounds where there is destructive interference

How do we explain interference patterns?

- The diagram shows the double slit experiment producing an interference pattern with light.

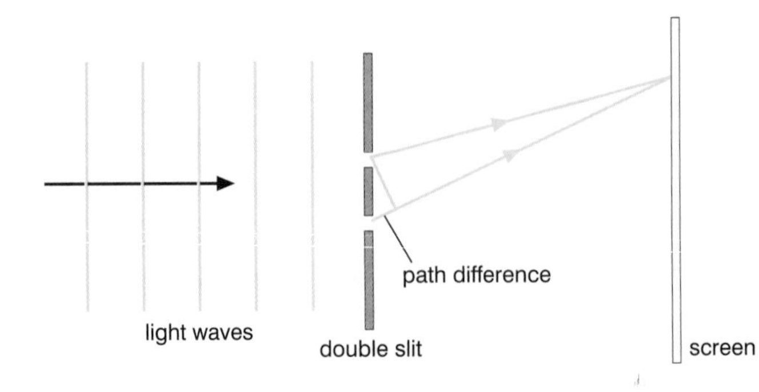

QUESTION SPOTTER

- The example is about light, but the results are exactly the same for water waves, microwaves or sound.

- The two sets of waves travel different **distances** to reach any point on the screen, except for the very centre point. This means there is a path difference between the two sets of waves. The **path difference** is the **extra distance** that one wave has to travel to reach the screen.

- For *constructive interference*, the two sets of waves must arrive at exactly the same point during their cycles, so the path difference must be equal to a *whole number of wavelengths*.

- For *destructive interference*, the two sets of waves must arrive at exactly opposite points during their cycles, so the path difference must be equal to a whole number of wavelengths *plus an extra half a wavelength*.

A* EXTRA

- The technical term for 'waves arrive at the same point in their cycle' (constructive interference) is to say that the waves arrive *in phase*. For destructive interference, the waves need to arrive *in antiphase*, i.e. they are at opposite points in the cycle.

IDEAS AND EVIDENCE

In the 1600s, there were two theories about how light moves. Newton favoured the idea that light moves as a stream of particles – he called them corpuscles. Huygens thought that light moves as a wave, like ripples in a pond. At that time, there were no experiments that could tell which was the correct idea. It was only in the early 1700s that Thomas Young managed to show interference effects in light. This evidence supported the wave idea of Huygens.

Check Yourself Questions

Q1 Describe an experiment to show interference of waves in a school lab.

Q2 A helicopter pilot has a special helmet. A microphone picks up the noise from the engine and an electric circuit quickly turns this wave upside down before feeding the signal into speakers in the pilot's helmet. Use ideas about interference to explain how this reduces the noise the pilot hears.

Q3 Look at the Ideas and Evidence box again. Explain why experimental work is important in science.

Answers are on page 147.

REVISION SESSION 1

Information transfer and storage

The history of communication

- Throughout human history, there has been progress in the **speed** of transferring information and in the **distances** covered. Some significant steps were:

  1793 – Claude Chappe invented the first long-distance semaphore system. This meant that messages could be sent from place to place by light – much faster than sending a runner or a rider. It is restricted to 'line-of-sight', however, so each step in the chain is limited by the need to see the next station.

  1843 – Samuel Morse invented the first long-distance telegraph line. Using an electrical signal in a wire, messages could now be sent over much greater distances at one go, increasing the speed of information transfer. This is restricted by the need to have a cable connection between the sender and receiver.

  1876 – Alexander Graham Bell patented his design for the telephone.

  1888 – Heinrich Hertz first demonstrated electromagnetic waves – the discovery that led directly to radio, television and radar.

  1902 – Marconi transmitted a radio signal across the Atlantic Ocean. Signal transfer can now be wireless.

  1924 – Edward Appleton proved the existence of the ionosphere, the atmospheric layer responsible for reflecting sky waves back to Earth.

  1927 – The first television broadcasts were made in Britain. Pictures could now be transmitted as well as sound.

  1962 – Telstar was the first satellite to transmit television pictures.

  By **2004**, information could be transmitted via telephone, television, fax, email and the internet.

IDEAS AND EVIDENCE

You do not need to learn the exact dates and events. The important thing is to remember that, at each stage, the keywords are: more, faster and further. Each major advance can be linked to a development in ideas or in the technology available – the use of electrical cables, the use of radio waves, the use of fibre optic cables.

What is a communications system?

- In a communications system, some information – the **signal** – is **encoded** in some way, fed into a **communications link** – which might be electrical or optical – and then **decoded** to reproduce the original signal.

- These steps can be broken down further, to give a system with a number of separate blocks, each performing a specific task.

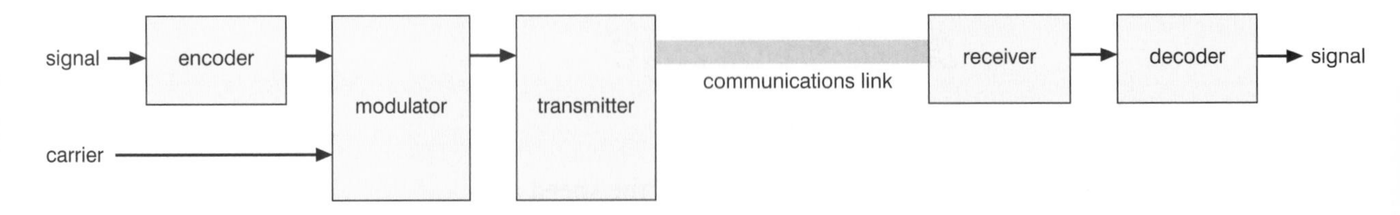

- The **encoder** transforms the signal to an agreed code. This signal is used to modulate the carrier in the **modulator** (see Revision Session 2). The **transmitter** sends the signal. The signal arrives at the **receiver** and the **decoder** retrieves the signal. There may be **storage devices** at the sending end or the receiving end, and an **amplifier** may be needed to increase the amplitude of the signal.
- The **storage devices** could be CDs, floppy discs, computer hard discs or magnetic cassette tape.
- A **transducer** is any device that transfers energy from one form to another – for example, loudspeakers (electrical to sound), microphones (sound to electrical) and record/playback heads in tape recorders (see Unit 2, Revision Session 3).

Analogue and digital

- In **analogue** coding, the processed signal can vary over a range of values. The pattern produced follows the original information.
- In **digital** coding, the processed signal can only take only two values.

- Questions on the difference between digital and analogue signals are common. Be clear and concise in your answer.

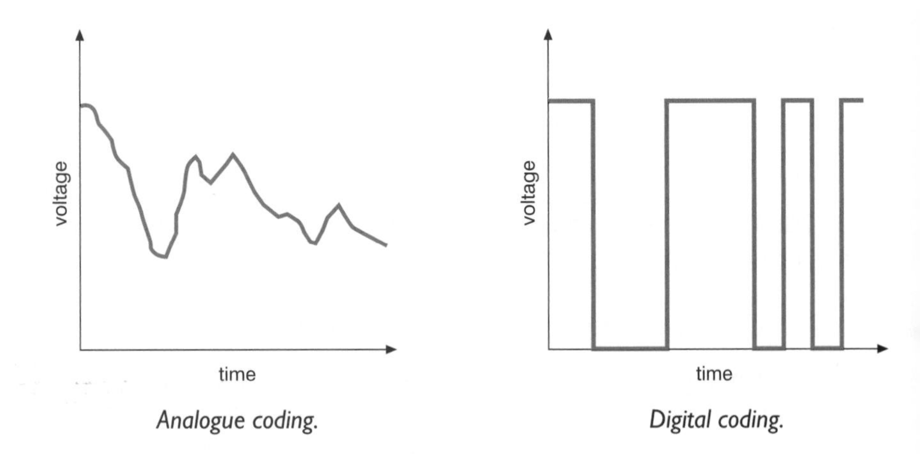

Analogue coding.

Digital coding.

Attenuation and noise

- Three important methods of sending a signal are by using **radio waves**, **light** and **infrared**. Light and infrared can be sent along optical fibres using **total internal reflection** (see Unit 6, page 83). Radio systems are covered in the next Revision Session.

- When a signal is sent along a cable or optical fibre, energy losses occur that mean the strength of the signal becomes weaker. This is called **attenuation**. Also, unwanted signals called **noise** are added in to the main signal. Noise can be generated by, for example, interference from other radio signals. **Repeaters** or **regenerators** are used to compensate for these losses.

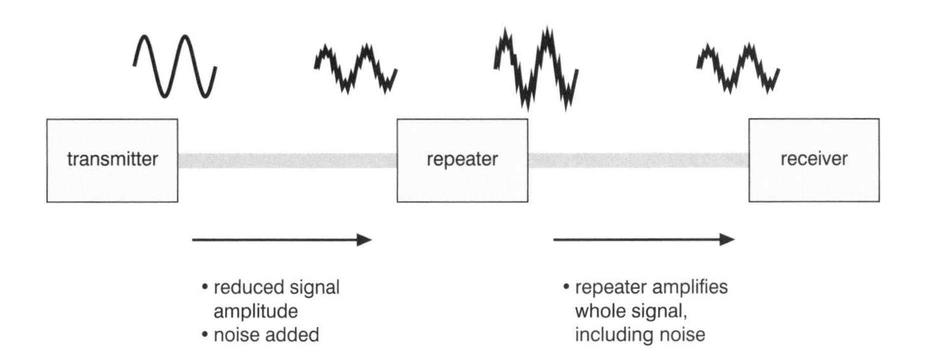

- Repeaters amplify the signal they receive. They cannot distinguish between the original signal and the noise, so they amplify both – this is a disadvantage of analogue systems.

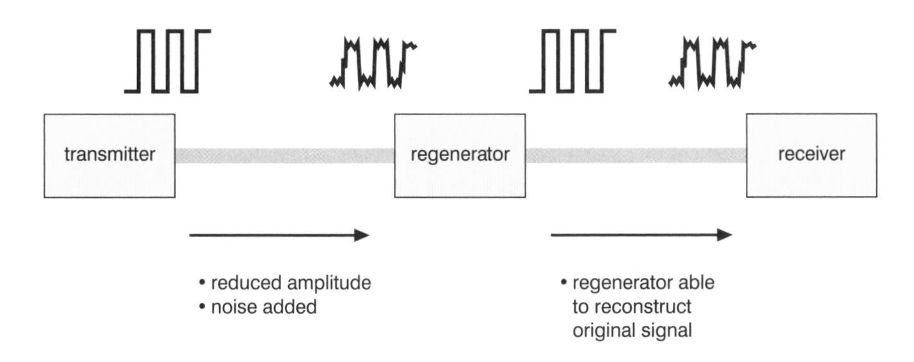

- With a digital signal, the regenerator only needs to recognise whether the signal is 'high' or 'low', so the original signal can be reconstructed.

CHECK YOURSELF QUESTIONS

Q1 Billy sends a message to Holly using morse code using a torch. In this system:
a What is the modulator?
b What is the receiver?
c What is the decoder?

Q2 Jenny owns a record player and a CD player. She notices that a scratched CD can play normally, but when a record has a scratch on it, the scratch is always heard in the sound from the speakers. Use ideas about analogue and digital signals to explain why this happens.

Answers are on page 147.

Radio systems

How do radio waves travel?

- Radio waves are part of the **electromagnetic spectrum** (see Unit 6). They travel at the speed of light and their speeds, frequencies and wavelengths are linked by the wave equation, $v = f\lambda$. **TV signals** form part of the radio spectrum.
- Radio waves used for communication travel to the receiver in three ways.
  - **Ground waves** follow the surface of the Earth and are used for frequencies up to 2 MHz. They have a range of about 1000 km.
  - **Sky waves** are reflected back to Earth by a layer of the atmosphere – the **ionosphere**. These waves have frequencies between 3 and 30 MHz.
  - Frequencies higher than 30 MHz travel through the ionosphere and are reflected back to Earth by satellites. These waves are called **space waves**.

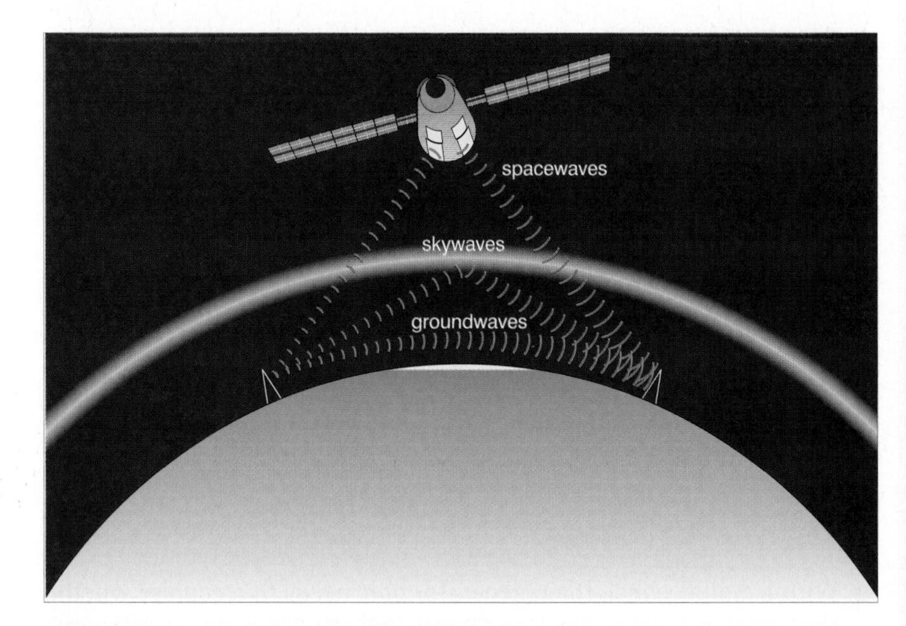

How radio waves travel from the transmitter to the receiver.

A* EXTRA

- Interference effects can be explained using ideas about path difference (see Unit 6, page 90).

- **Diffraction** is important in radio communication. It allows the ground waves to follow the curvature of the Earth and accounts for radio waves 'spreading' into valleys. Diffraction effects happen most strongly when the wavelength of the wave is close to the size of the gap involved. Radio signals have wavelengths up to 2 km and beyond, which diffract well through features of the landscape.
- **Interference** effects occur when a signal is received after following two different routes. For example, a signal might be received directly as well as being reflected from a nearby cloud or building. When this happens, the two signals produce a resultant wave which may fade in and out, or which fades and reappears if it is a TV signal. Interference effects can also happen in a car when signals arrive from two different transmitters. The resultant signal can vary in strength as the car moves between the two transmitters.

How is information coded in radio signals?

- Two methods for coding information in radio signals are **amplitude modulation (a.m.)** and **frequency modulation (f.m.)**. Both systems use a **carrier wave** with a high frequency and then add the required signal to it. At the receiving end, the carrier wave is removed to leave the required signal.
- With **amplitude modulation**, the two signals are combined to send a signal with a variable amplitude.

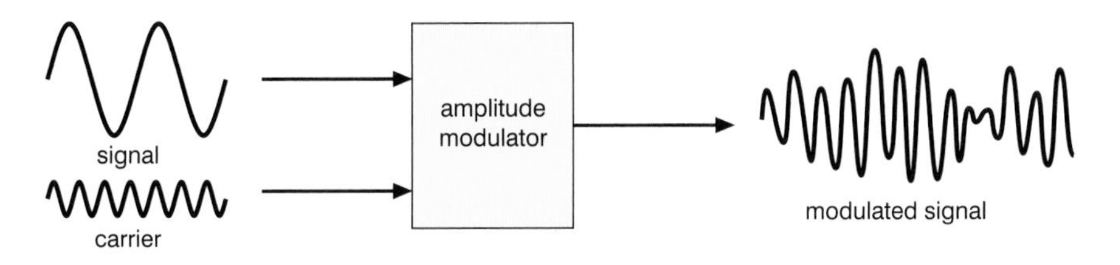

- Advantages of this system are that the signal can be sent over a greater distance and a smaller range of frequencies is needed to transfer the information. A disadvantage is that this kind of signal picks up more noise, which is then amplified together with the signal.
- With **frequency modulation**, the two signals are combined to send a signal with a variable frequency.

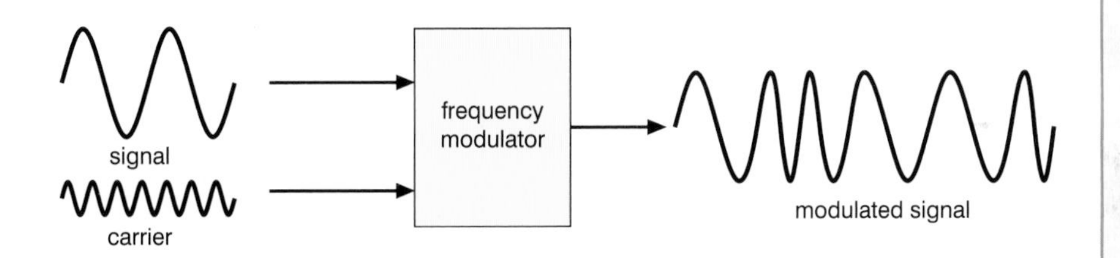

- Advantages of this system are better sound quality and less noise being amplified. However, a disadvantage is that the signal can only be sent over a limited distance.

IDEAS AND EVIDENCE

The 'digital revolution' has made huge amounts of information available to millions of people. This has brought many benefits, such as being able to shop via the internet and being able to keep in touch using mobile phones. However, it has also brought potential hazards, including internet fraud and concerns over the long-term safety of mobile phones. Be prepared to discuss these issues and suggest how science might, or might not, be able to solve some of these problems.

Using satellites

- Space wave radio signals use satellites to reach a wider footprint. (The footprint is the part of the surface of the Earth that can receive the signal.)
- Satellites can be:
  - **passive** – they reflect the signal straight back to Earth –

  or
  - **active** –the satellite receives the signal, processes it and then re-transmits it back to Earth.

  For more about communications satellites, see Unit 8.

CHECK YOURSELF QUESTIONS

Q1 Radio waves travel at the speed of light, 3×10^8 m/s. Complete the table.

Radio station	Frequency	Wavelength
BBC Radio 2	89.9 MHz	
BBC Radio 4		198 m
BBC Radio 5 Live		909 m

Q2 Describe the difference between a repeater and a regenerator. What effect does this difference have on the signal transmitted?

Q3 Reece lives in a valley. He receives radio stations clearly but receives a poor television signal, even though the signal is broadcast in his area.

a Use diffraction ideas to suggest why Reece's reception is different for radio and TV signals.

b Use interference ideas to suggest a different reason for Reece's poor television reception.

Answers are on page 147.

REVISION SESSION 1 The Solar System

What makes up the Solar System?

- The **Solar System** is made up of the Sun and its planets. The planets are **satellites** of the Sun and are kept in **orbit** by the gravitational pull from the Sun. The orbits of the planets are slightly oval or elliptical.

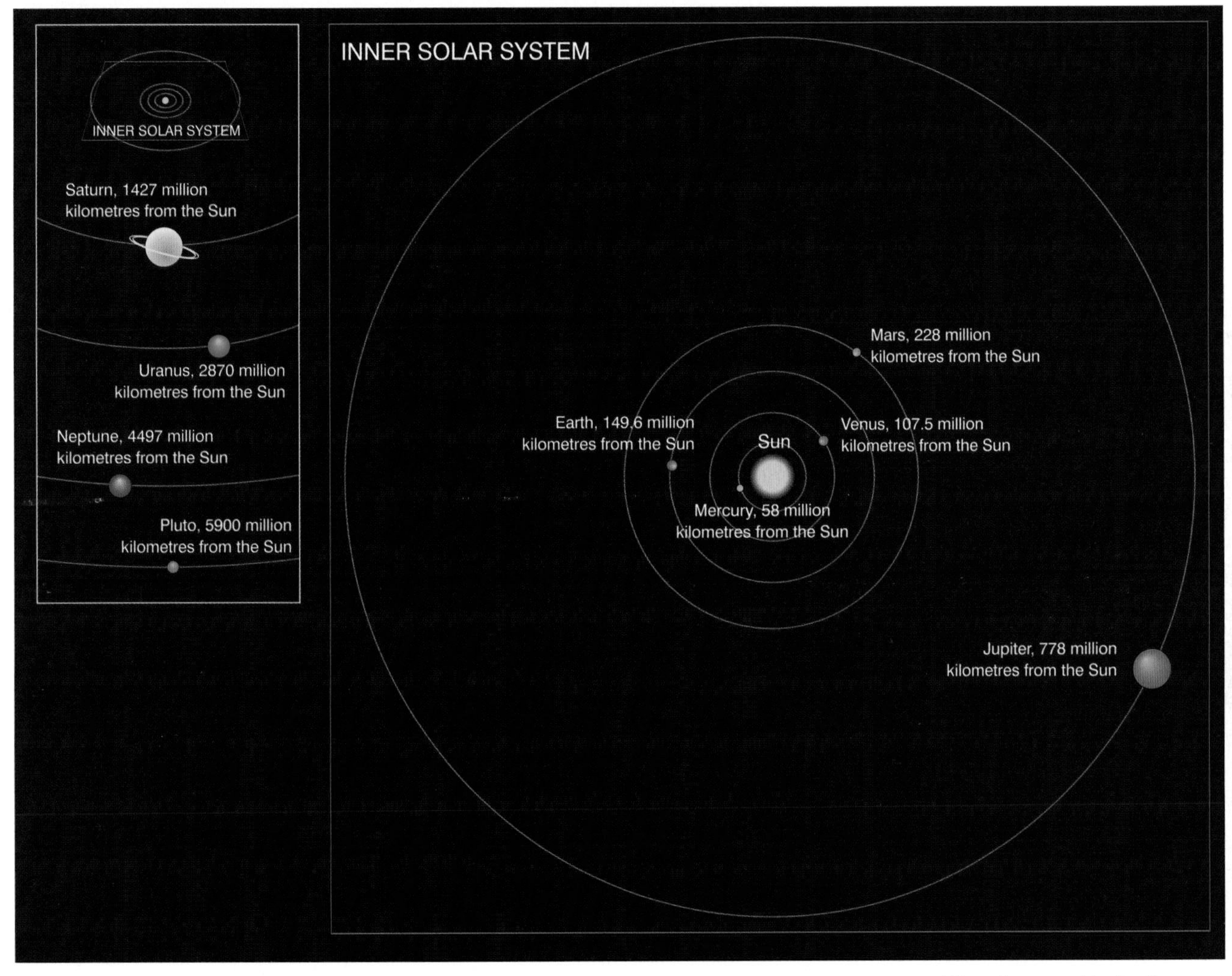

Nine planets orbit the Sun.

- **Asteroids** are fragments of rock, up to 1000 km in diameter, that orbit the Sun between the four inner planets and the five outer planets. The asteroids were formed at the same time as the Solar System.

- **Comets** go round the Sun like planets but their orbits are much more elliptical. Comets spend much of their time too far away from the Sun to be seen. They are thought to be made from ice and rock. When they get close to the Sun some of the solid turns into gas, forming a 'tail' which points away from the Sun.

A comet moves round the Sun in an elliptical orbit.

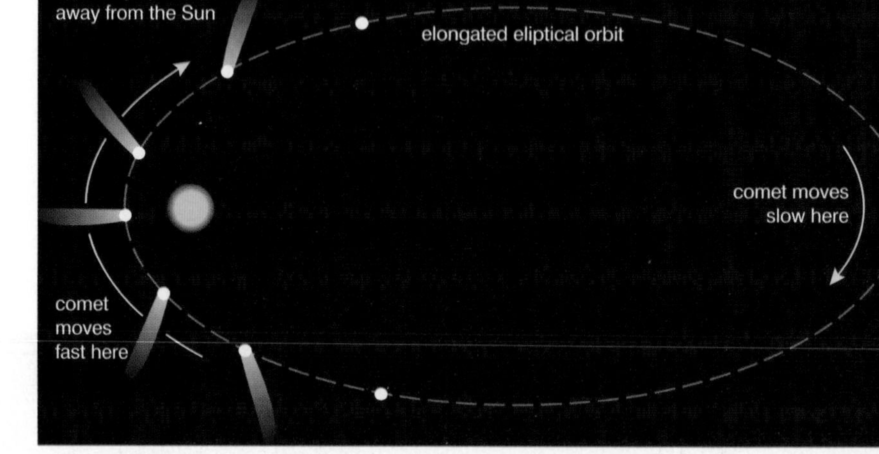

A* EXTRA

- Rockets carry their own fuel and also carry oxygen to react with the fuel. (There is no oxygen in space!) When the fuel reacts with the oxygen it forces the exhaust gases backwards – this forces the rocket forwards.

Satellites

- Near the Earth, gravity exerts a force that always acts towards its centre and keeps satellites like the Moon moving in a near-circular orbit around the Earth. Artificial satellites orbit the Earth for the same reason – and as they do not need motors or rockets to stay in such orbits, they can stay there for decades.
- Artificial satellites in orbit have many uses in communication (see Unit 7 Revision Session 2). For example, they can be used to transfer TV signals, survey the Earth's surface or for military uses (spy satellites).
- At the right distance from the Earth, an unpowered orbit takes exactly 24 hours. This means that an artificial satellite orbiting above the Earth's equator and moving in the same direction as the Earth's spin, will appear to be flying over the same spot. This is a **geostationary** orbit and is very important in satellite communications. It means that a transmitter or a receiver can be pointed permanently at the same point in the sky, without the need to track the motion of the satellite in its orbit.
- Another important orbit for satellites is a **polar** orbit. The satellite follows an orbit that crosses the North and South poles of the Earth. As the Earth is spinning, this means that the satellite goes over a different part of the surface each time it orbits. This is useful in geographical surveys.

Centripetal force

- Natural and artificial satellites orbit the Earth in a circle for the same reason that the planets orbit the Sun – **gravitational pull**.
- To keep an object moving in a circle, you have to constantly pull it towards the centre of that circle. A force that constantly pulls towards a centre is called a **centripetal force**.

- The **size** of the force involved can be calculated using this formula:

$F = \frac{mv^2}{r}$

F = force in Newtons
m = mass of object in kilograms (kg)
v = speed of object in metres per second (m/s)
r = radius of orbit in metres (m)

- The **speed** of an object in circular motion is calculated using the usual formula:

$\text{speed} = \frac{\text{distance}}{\text{time}}$

- For objects in circular motion, we often refer to the **time period**, which is the time taken for one complete orbit. In this situation:

$\text{speed}, v = \frac{\text{distance around the circle}}{\text{time period}} = \frac{2\pi r}{T}$

r = radius of circle in metres (m)
T = time period in seconds (s)

A* EXTRA

- An object in orbit is constantly changing direction – its *velocity* is changing. If an object is changing velocity, it is *accelerating*. If an object is accelerating, there must be a *force* to cause the acceleration. In circular motion, this force is *always* towards the centre of the circle.

WORKED EXAMPLE

A survey satellite orbits the Earth 200 km above the surface. It has a mass of 100 kg and orbits once every 90 minutes. Calculate the centripetal force on this satellite. (The radius of Earth is 6400 km.)

First, work out the speed of the satellite.
Write down the formula for speed: $\text{speed} = \frac{\text{distance around one orbit}}{\text{time period}}$

Substitute the values for distance and time: $\text{speed} = \frac{(2 \times \pi \times 6\,600\,000)}{5400}$

(radius of orbit = radius of Earth + 200 km = 6600 km = 6 600 000 m, time period = 90 min = 5400 s)

Work out the answer and write down the unit: speed = 7700 m/s

Now, work out the force on the satellite.
Write down the formula for force: $F = \frac{mv^2}{r}$

Substitute the values for m, v and r $F = \frac{(100 \times 7700^2)}{6\,600\,000}$

Work out the answer and write down the unit: F = 898 N

Galaxies

- The Sun is part of a group of stars called the Milky Way **galaxy**. The Milky Way is in the shape of a spiral.
- We can see other galaxies through telescopes. These other galaxies are so far away that it is difficult to see the individual stars, so earlier astronomers called them **nebulae** (singular: nebula) meaning 'a bright cloudy spot'. These days, we can see these galaxies properly, and the word nebula now usually means a cloud of gas where stars are being born. Many nebulae are formed by stars exploding.

Life cycle of a star

Birth (thousands of millions of years)	Life (tens of millions of years)	Death (millions of years)
• The star starts as a huge cloud of hydrogen gas and dust. • Gravity pulls the hydrogen atoms closer together. • The gas gets hotter as the cloud gets smaller and more concentrated. • In the very high temperatures created, the hydrogen atoms join together to form helium atoms in a process called nuclear fusion. Large amounts of energy are released in this process. • The core sends out light and other radiation and the cool, outer layers are blown away, sometimes forming planets around the new star.	• Huge gravitational forces pull the outer parts of the star towards its core. • The energy produced by the nuclear fusion ensures that the gases on the outside are very hot. • The pressure exerted by the hot gases exactly balances the force of gravity. • This balance continues for millions of years until all the hydrogen has been used up and the nuclear reactions stop. • Some stars appear red-orange but hotter stars appear blue-white.	• The hydrogen in the star's core gets used up and the core starts to cool. • The star then starts to collapse and the core gets hotter again very quickly, enabling other nuclear fusion reactions to start. • A sudden surge of radiation is produced and the star expands to form a red giant or red supergiant. • In a medium-sized star like the Sun, the nuclear fusion reactions finish and the core collapses under gravity and the red giant forms a white dwarf star. • In a large star, when the core collapses there is a sudden explosion and the red supergiant forms a supernova. A supernova is brighter than a whole galaxy of stars. A very dense core is left behind becoming a neutron star and, if it is very small and dense, a black hole.

- Questions on the life cycle of a star will usually require an extended written answer and be worth 3 marks or more. Divide your answer into three parts: birth, life and death. Remember that what happens at the death of the star will depend on its size.

IDEAS AND EVIDENCE

NASA have recently announced the discovery of Sedna, a planet-like body three times further from the Sun than Pluto. With a diameter of 800–1100 miles, Sedna is about three-quarters of the size of Pluto (diameter 1400 miles) and half the diameter of our Moon (diameter 2100 miles). Sedna is the largest known object to orbit the Sun beyond the orbit of Pluto.

CHECK YOURSELF QUESTIONS

Q1 How does a normal star produce energy?

Q2 The list contains the stages in the life cycle of a star. Arrange them in the correct order.

supernova *clouds of hydrogen gas*
red supergiant *blue star*

Q3 **a** A satellite moves around the Earth at a constant speed in a circular orbit.
i What force is acting on the satellite?
ii What is the direction of the force?

b Some satellites travel in geostationary orbits. Explain what is meant by a geostationary orbit.

Answers are on page 148.

How did the Universe begin?

Red shift

- If a star is moving away from the Earth, light waves will reach the Earth with a lower frequency than that emitted by the star. This change in frequency is called a **red shift**.
- Light from all the distant galaxies shows a red shift, so it follows that all these galaxies are moving away from us. In fact, **all galaxies are moving away from each other.** The Universe is expanding.

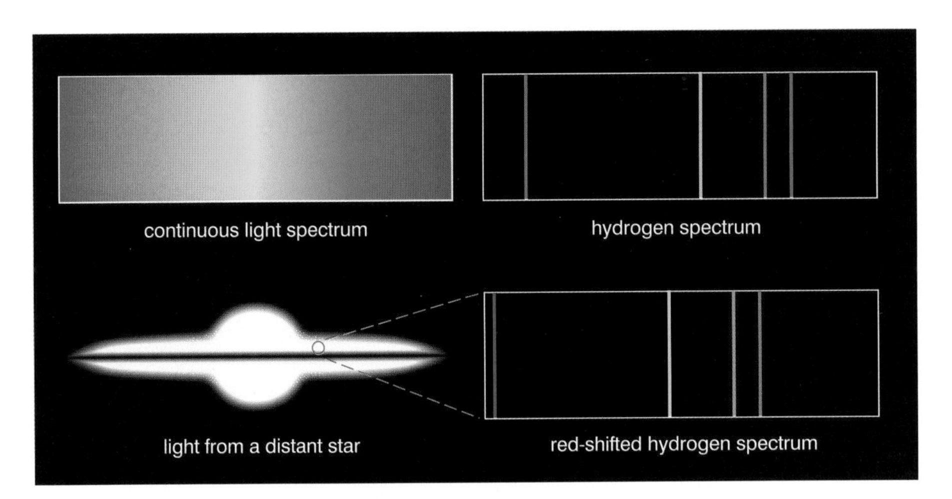

The top picture shows the spectrum of white light. The second picture shows the spectrum produced by hydrogen when it glows. The bottom picture shows the spectrum of hydrogen from a star seen through a telescope. The pattern is shifted towards the red end. This red shift shows that the star is moving away from us.

A* EXTRA

- One of the pieces of evidence that suggests the universe is expanding comes from examination of the emission spectra of hydrogen from distant stars. The lines in the spectra are at lower frequency than on Earth. This shift towards the red end of the spectrum (red shift) indicates that the distant stars are moving away from the Earth.

The Big Bang theory

One explanation for the Universe expanding is that everything in the Universe originated from the same point, which exploded in the **Big Bang**. Since this explosion, the Universe has been expanding and cooling. Calculations suggest that the Big Bang could have occurred between 11 and 18 billion years ago. There are three main possibilities as to what will happen to the Universe:

1. **The Universe will keep on expanding**. The galaxies move fast enough to overcome the forces of gravity acting between them.
2. **The expansion will slow down and stop.** The galaxies will remain in a fixed position.
3. **The Universe will start to contract**. The force of gravity between the galaxies will overcome the forces causing expansion and pull them back together again.

QUESTION SPOTTER

- Questions on the future of the Universe will usually require an extended answer. You should refer to the three possibilities and mention the importance of the differences between the competing forces acting on the galaxies.

Looking for life elsewhere

- Scientists looking for life in the universe look for evidence of:
  - suitable **chemicals** – in particular carbon, water and oxygen
  - suitable **temperatures** – life needs liquid water, so this limits the possible temperature range
  - suitable **places** – in recent years scientists have discovered a number of planets which circle other stars apart from the Sun. Most of these planets are large gas giants, like Jupiter.
- The methods scientists use to gather this evidence include:
  - direct evidence from rocks **gathered** on space missions – brought back from the Moon or studied on the surface of Mars by robots
  - direct evidence from meteorites and other material from space that has **landed** on the Earth
  - using telescopes to study the **light** from objects in the sky – different chemicals produce characteristic spectral colours
  - using radio telescopes to look for evidence of **signals** from space – the SETI@home project allows people to analyse sections of data on their home computers.

IDEAS AND EVIDENCE

This whole section is about interpreting a limited amount of data – scientists cannot travel all over the universe to conduct experiments! Ideas change as new evidence is gathered, sometimes because a new technique is invented, such as using radio waves for observation as well as visible light. Sometimes the theory suggests new ideas that experimenters can use to guide their observations, such as searching for black holes.

CHECK YOURSELF QUESTIONS

Q1 What is the red shift?

Q2 What is the evidence that supports the Big Bang theory?

Answers are on page 148.

REVISION SESSION 1

Kinetic theory of gases

How do gases behave?

- Experimentally, gases follow these simple rules:
  - If the pressure of the gas stays constant, then the **volume** of the gas is proportional to its **temperature**. **(Charles' Law)**
  - If the volume of the gas stays constant, then the **pressure** of the gas is proportional to its **temperature**. **(Pressure Law)**
  - If the temperature of the gas stays constant, then the **volume** of the gas is **inversely** proportional to the **pressure**. **(Boyle's Law)**
- However, gases *only* follow these rules if two conditions are met:

  1 The mass of the gas must remain constant (i.e. no particles move in or out of the system).

  2 The temperature must be measured using the **Kelvin** scale.

What is the Kelvin scale?

- These graphs show typical results for volume–temperature and pressure–temperature experiments.

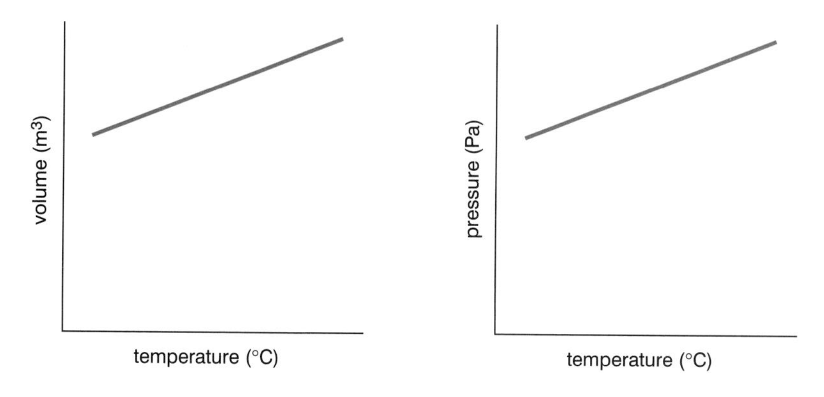

- If the lines are continued to lower temperatures, they reach the temperature axis at -273 °C. This temperature is called **absolute zero**.

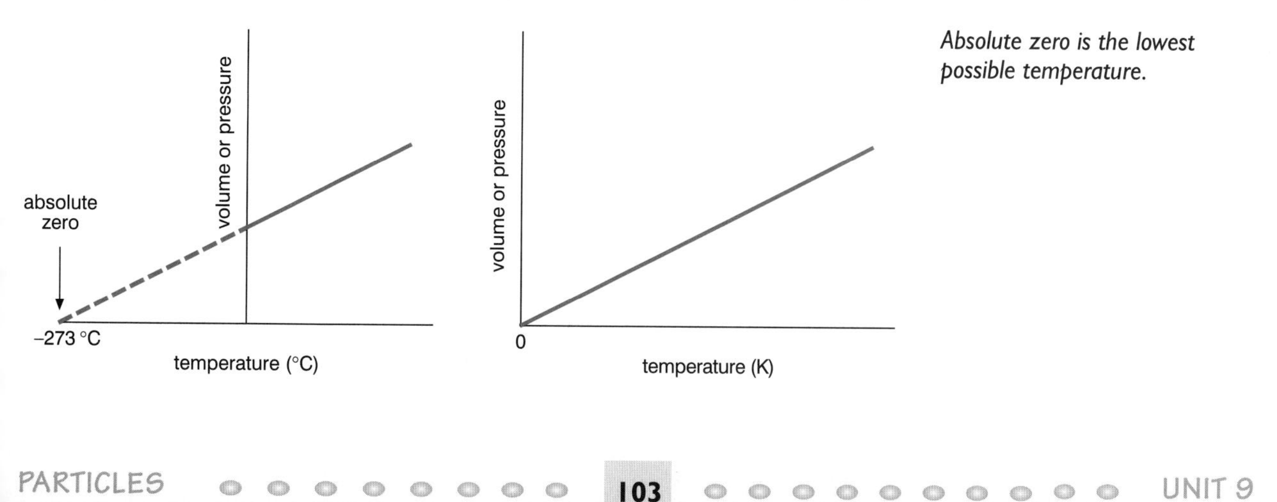

Absolute zero is the lowest possible temperature.

- Notice that temperatures are measured in 'Kelvin' and not in 'degrees Kelvin'.

- The Kelvin scale of temperature uses absolute zero as its starting point and counts upwards from there. To make conversion to °C easier, the steps along both scales are the same.

  To make the conversion:

  temperature in °C = temperature in Kelvin – 273
  temperature in Kelvin = temperature in °C + 273

Example	Temperature in °C	Temperature in Kelvin
Absolute zero	–273	0
Freezing point of water	0	273
Room temperature	20	293
Boiling point of water	100	373

Using gas law equations

- The experimental rules about how gases behave can be summarised as equations. Usually, we are interested in what happens during a change, so there are usually 'before' and 'after' situations.
- The most general equation is:

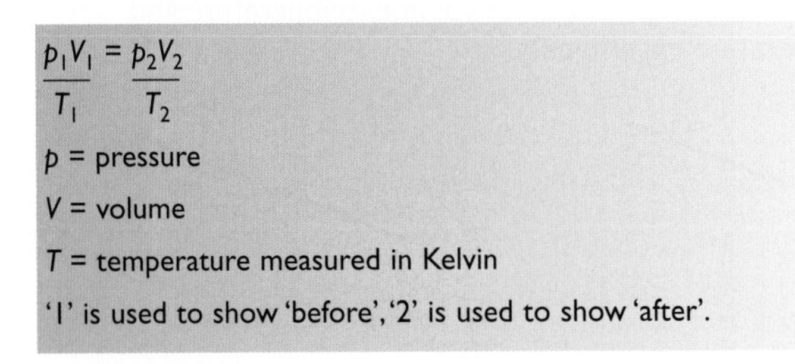

$$\frac{p_1V_1}{T_1} = \frac{p_2V_2}{T_2}$$

p = pressure

V = volume

T = temperature measured in Kelvin

'1' is used to show 'before', '2' is used to show 'after'.

A* EXTRA

- The units of volume and pressure need to be consistent before and after any changes, but the temperature *must* be in Kelvin.

- Often, the equation can be simplified because one of the variables remains constant, e.g. the temperature does not change.

WORKED EXAMPLE

At 10 °C, the pressure in a car tyre is three times atmospheric pressure. Assuming the volume of the tyre does not change, what will be the pressure at 40 °C?

Write down the equation: $\frac{p_1V_1}{T_1} = \frac{p_2V_2}{T_2}$

The volume does not change, so the equation becomes: $\frac{p_1}{T_1} = \frac{p_2}{T_2}$

Rewrite the formula in terms of p_2: $p_2 = \frac{p_1T_2}{T_1}$

Substitute the values for p_1, T_1 and T_2, remembering to change the temperatures to Kelvin (10 °C = 283 K, 40 °C = 313 K):

$$p_2 = \frac{3 \times 313}{283}$$

$$= 3.3$$

So final pressure = 3.3 times atmospheric pressure.

The kinetic theory of gases

- In the kinetic theory of gases, we build up a set of ideas from the basic idea that a gas is made of many tiny particles, called **molecules**. These ideas give us a picture of what happens inside a gas.

Observed feature of a gas	Related ideas from the kinetic theory
Gases have a mass that can be measured.	The total mass of a gas is the sum of the masses of the individual molecules.
Gases have a temperature that can be measured.	The individual molecules are always moving. The faster they move (the more **kinetic energy** they have), the higher the temperature of the gas.
Gases have a pressure that can be measured	When the molecules hit the walls of the container they exert a force on it. It is this force, divided by the surface area of the container, that we observe when measuring pressure.
Gases have a volume that can be measured	Although the volume of each molecule is only tiny, they are always moving about and spread out throughout the container.
Temperature has an absolute zero.	As temperature falls lower, the speed of the molecules (and their kinetic energy) becomes less. At absolute zero the molecules would have stopped moving.

The kinetic theory of gases is a very good example of a theory in action – starting with observed evidence, scientists construct a mathematical and mental model of an invisible world of molecules in motion. By following these ideas and using knowledge from other parts of science, for example what happens during collisions, scientists produce a series of predictions, which can then be tested in further experiments.

- These ideas help to explain the three gas laws mentioned above:

Experimental gas law	Link to the kinetic theory
If the pressure of the gas stays constant, then the volume of the gas is proportional to its temperature. (Charles' Law)	A higher temperature means the molecules move more quickly, so the force on the walls will be higher. If the pressure stays constant (pressure = force / area) and the force is higher, then the volume must increase to give a larger surface area.
If the volume of the gas stays constant, then the pressure of the gas is proportional to its temperature. (Pressure Law)	A higher temperature means the molecules move more quickly, so the force on the walls will be higher. If the volume of the gas is constant (which means the surface area will stay constant), then the pressure (= force / area) must increase.
If the temperature of the gas stays constant, then the volume of the gas is **inversely** proportional to the pressure. (Boyle's Law)	The temperature stays constant, so the average speed of the molecules stays constant. If the volume of the gas is reduced by half, then the molecules make the same number of collisions with half the surface area of wall, so the pressure (= force / area) must be doubled. This is **inverse** proportionality.

Check Yourself Questions

Q1 How does kinetic theory explain the idea of absolute zero?

Q2 **a** Convert these temperatures from °C to Kelvin.

i 20 °C **ii** 150 °C **iii** 1000 °C

b Convert these temperatures from Kelvin to °C.

i 300 K **ii** 650 K **iii** 1000 K

Q3 Cassie blows up a balloon. At room temperature, 20 °C, she measures the volume of the balloon as 1500 cm^3. Then Cassie puts the balloon in a freezer where the temperature is –13 °C. Assuming the pressure stays constant, work out the new volume of the balloon.

Answers are on page 148.

Developing ideas about atoms

Early ideas

- The first ideas about atoms were suggested in about 450 BC by the Greeks, Leucippus and Democritus. Their ideas were not widely accepted and, because there was no experimental work to support the idea, the argument could not be settled.
- The idea of atoms as small particles that could not be broken into anything smaller was taken up by a number of people over the following centuries. For example, Newton made use of the idea of 'corpuscles' in his work.

And so to the 1800s

- Our modern ideas of atoms are based on the work of John Dalton, announced in 1803. He made use of experimental work in chemistry, particularly the work of Lavoisier, to propose a system of separate atoms that could combine to form different chemical compounds.
- In the 1890s, J.J. Thomson was experimenting with an effect called cathode rays (see Revision Session 6). He proposed that these were made from a stream of particles, which were smaller than atoms. He called these particles **electrons**.
- This led to the development of the '**plum pudding**' model of the atom. In the plum pudding atom, the solid atom had tiny electrons spread through it, like the plums in a plum pudding.

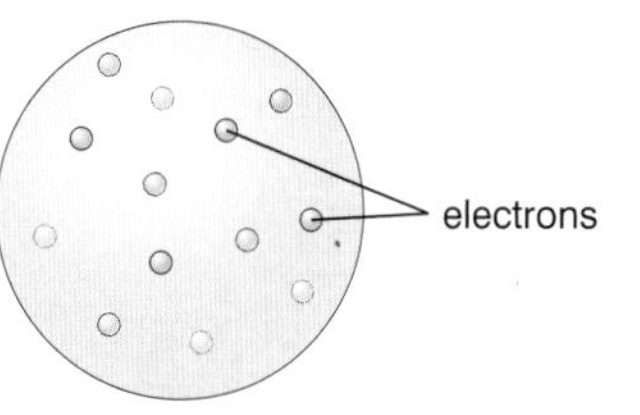

The plum pudding atom.

How was the 'nuclear' atom discovered?

- In a crucial experiment in 1911, a stream of alpha particles was aimed at a piece of gold foil a few atoms thick. Most of the alpha particles passed through the foil undeflected, showing that most of the atom is empty space. Some alpha particles passed through but were deflected and a few bounced back as if they had done a U-turn.

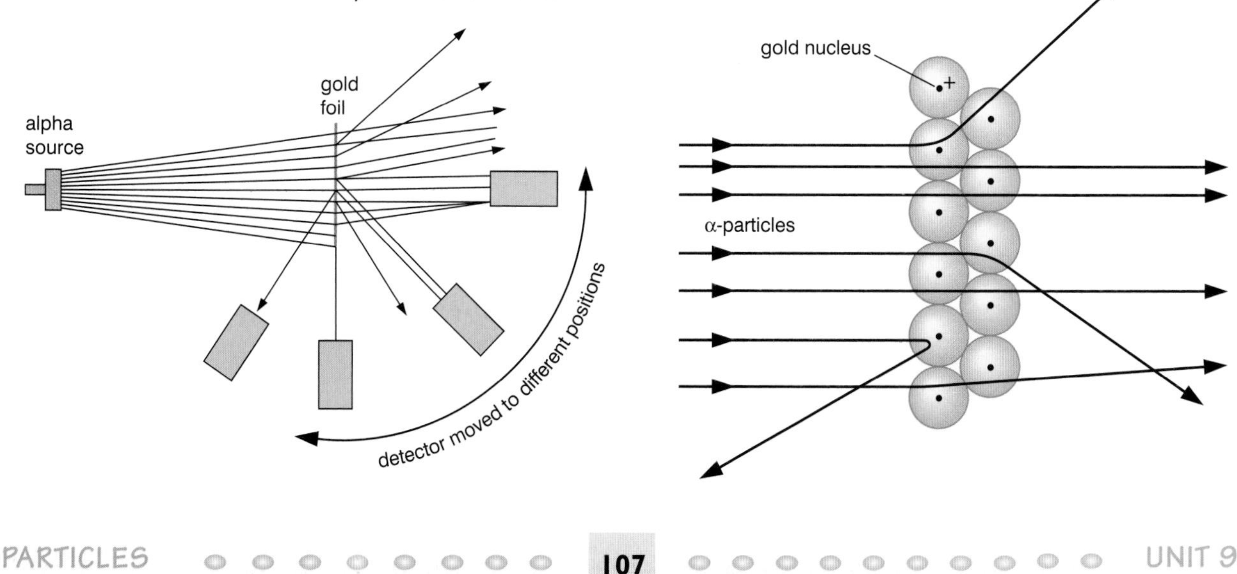

A* EXTRA

- The expected result was that all the alpha particles should go straight through. This would confirm the average density of the atoms. Only by concentrating the mass at the centre (in a nucleus) could the matter be sufficiently dense to deflect the alpha particles so much.

- The deflections meant that the atom's mass and charge is not spread out but must be clustered together.
- The very small number of U-turns must have been caused by 'head-on' collisions with a concentration of mass and positive charge. This meant that such clusters were very small.
- Altogether, this showed that an atom has a tiny cluster of positive charge – a **nucleus**.
- Further experiments in the 1920s and 1930s confirmed that the nucleus itself was made of two smaller particles – protons and neutrons. This is the **Rutherford** model of the atom.

IDEAS AND EVIDENCE

In this experiment, it is the *anomalous* results that hold the key to what happens. If the experimenters had just looked for the expected results and not checked *all* angles, this important discovery would not have been made.

CHECK YOURSELF QUESTIONS

Q1 List two similarities and two differences between the plum pudding and Rutherford models of the atom.

Q2 The diagram shows some of the results of an experiment using some very thin gold foil that was carried out by the scientist Sir Ernest Rutherford.

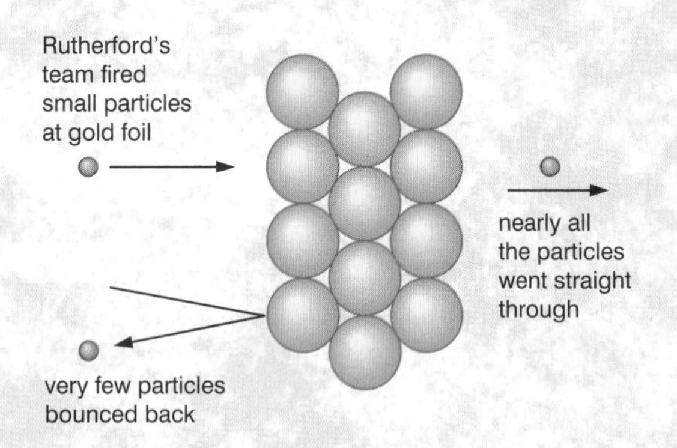

a What type of radioactive particle did Rutherford use?

b What did Rutherford deduce from the observation that most of the particles passed straight through the foil?

c What did Rutherford deduce from the observation that some particles bounced back from the foil?

Answers are on page 149.

Unstable atoms

Inside the atom

- Inside the atom the central **nucleus** of positively charged **protons** and neutral **neutrons** is surrounded by shells, or orbits, of **electrons.** Most nuclei are very stable, but some 'decay' and break apart into more stable nuclei. This breaking apart is called **radioactive decay**. Atoms whose nuclei do this are **radioactive**.
- When a nucleus decays it emits:
  - **alpha (α) particles**
  - **beta (β) particles**
  - **gamma (γ) rays.**
- A stream of these rays is referred to as **ionising radiation** (often called nuclear radiation, or just 'radiation' for short).

QUESTION SPOTTER

- You will often have to describe the differences between α, β and γ rays. You will be asked what the rays are, what materials they can penetrate and if they can be deflected.

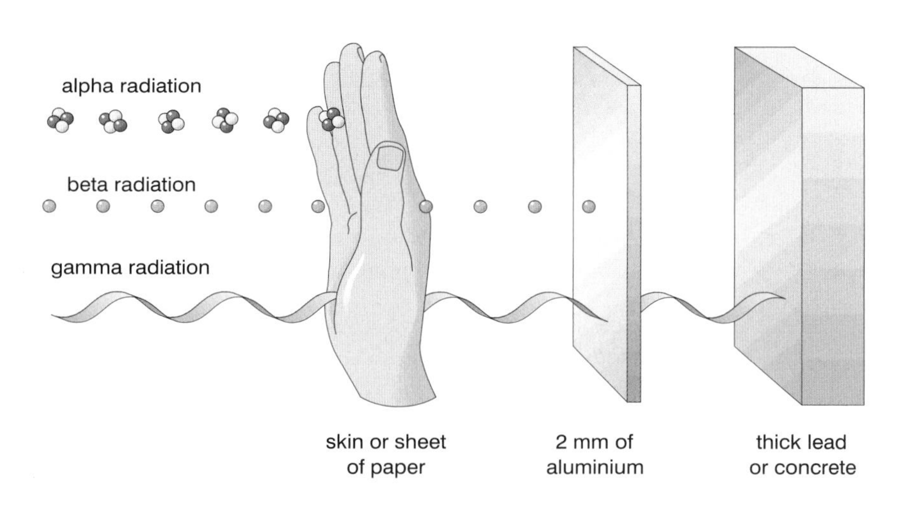

	alpha (α)	beta (β)	gamma (γ)
Description	A positively charged particle, identical to a helium nucleus (two protons and two neutrons)	A negatively charged particle, identical to an electron	Electromagnetic radiation. Uncharged
Penetration	4–10 cm of air. Stopped by a sheet of paper	About 1 m of air. Stopped by a few mm of aluminium	No limit in air. Stopped by several cm of lead or several metres of concrete
Effect of electric and magnetic fields	Deflected	Deflected considerably	Unaffected – not deflected

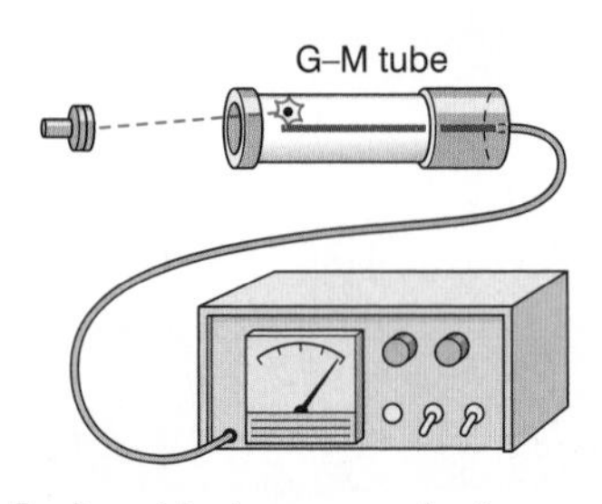

Radioactivity is measured using a Geiger–Müller tube linked to a counter.

Background radiation

- All ionising radiation is invisible to the naked eye, but it affects photographic plates. Individual particles of ionising radiation can be detected using a Geiger–Müller tube.
- There is *always* ionising radiation present. This is called **background radiation.** Background radiation is caused by radioactivity in soil, rocks and materials like concrete, radioactive gases in the atmosphere and cosmic rays from the Sun.

The N–Z curve

- Nuclei are formed in stars, where the conditions are extreme enough to force nuclear particles together. This is a random process and not all nuclei are equally **stable**. **Stability** is a measure of how long the nucleus is likely to stay together. For stable nuclei, a graph can be plotted linking *N*, the number of neutrons in a nucleus, with *Z*, the number of protons in a nucleus.

The N–Z *curve for stable nuclei.*

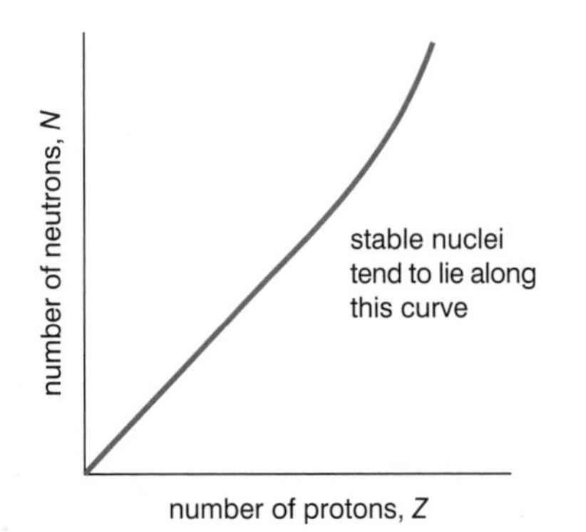

- If the unstable (i.e. radioactive) nuclei are added to the graph, they fall into distinct regions.

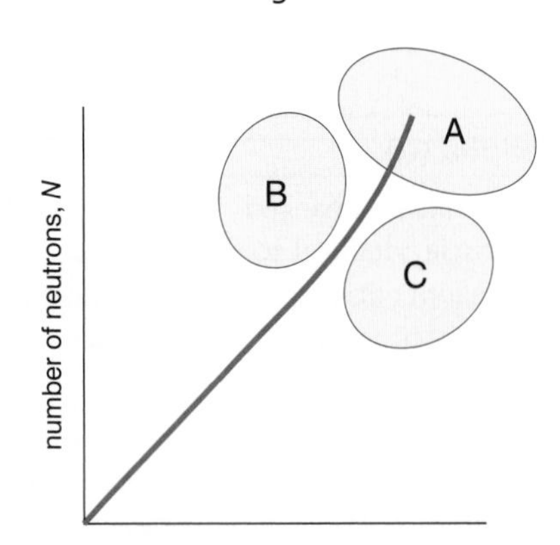

A* EXTRA

Nuclear reactions are very different from chemical reactions.
- The rate of the reaction (decay) does not depend on the concentration of the reactant.
- The time taken for half the atoms to decay is independent of how much reactant there is.
- The half-life is fixed for a given atom or isotope.

Region of graph	Nuclei in this region tend to be	Reason
A	alpha emitters	The nuclei are very large. Alpha emission reduces the mass of the nucleus by the largest amount (N reduces by 2, Z reduces by 2).
B	beta$^-$ emitters	Here the nuclei have a poor balance of protons and neutrons – too many neutrons. Beta$^-$ decay increases the number of protons (Z increases by 1) and reduces the number of neutrons (N reduces by 1).
C	beta$^+$ emitters	Here the nuclei have a poor balance of protons and neutrons – too many protons. Beta$^+$ decay reduces the number of protons (Z reduces by 1) and increases the number of neutrons (N increases by 1).
All three	gamma emitters	Particularly for beta emitters, the nucleus will have a surplus of energy after the emission of alpha or beta particles. This surplus energy is released as a burst of electromagnetic radiation – a gamma ray.

Half-life

- The **activity** of a radioactive source is the number of ionising particles it emits each second. Over time, fewer nuclei are left in the source to decay, so the activity drops. The time taken for half the radioactive atoms to decay is called the **half-life**.
- Starting with a pure sample of radioactive atoms, after one half-life half the atoms will have decayed. The remaining undecayed atoms still have the same chance of decaying as before, so after a second half-life half of the remaining atoms will have decayed. After two half-lives a quarter of the atoms will remain undecayed.

QUESTION SPOTTER

- Take care with questions on half-life. You may be asked about the *fraction remaining* after a particular time. After one half-life, $\frac{1}{2}$ of the original activity remains; after a second half-life, *half of that* remains, i.e. $\frac{1}{4}$; after a third half-life, *half of that* remains, i.e. $\frac{1}{8}$, and so on.

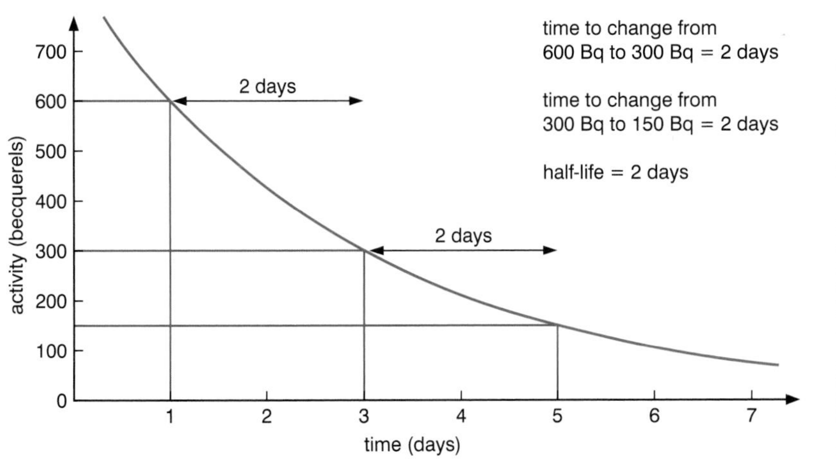

time = 0

time = 2 days

time = 4 days

The half-life is 2 days. Half the number of radioactive atoms decays in 2 days.

WORKED EXAMPLE

A radioactive element is detected by a Geiger–Müller tube and counter as having an activity of 400 counts per minute. Three hours later the count is 50 counts per minute. What is the half-life of the radioactive element?

Write down the activity and progressively halve it.
Each halving of the activity is one half-life:

0	400 counts
1 half-life	200 counts
2 half-lives	100 counts
3 half-lives	50 counts

3 hours therefore corresponds to 3 half-lives and 1 hour therefore corresponds to 1 half-life.

A* EXTRA

- In a nuclear equation an α particle is written as $^4_2\alpha$. This is because it is the same as a helium nucleus.
- A β particle is written $_{-1}^{0}\beta$ to show that it has negligible mass and is of exactly opposite charge to a proton.

Nuclear equations

- You can write down nuclear changes as **nuclear equations**. Each nucleus is represented by its chemical symbol with two extra numbers written before it. Here is the symbol for Radium-226:

the top number is the **mass number** (the total number of protons and neutrons)

the bottom number is the **atomic number** (the number of protons)

$$^{226}_{88}Ra$$

- The mass numbers and atomic numbers must balance on both sides of a nuclear equation.

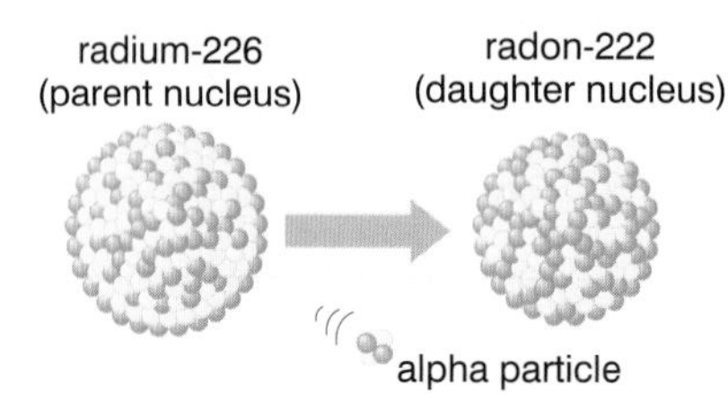

$$^{226}_{88}Ra \rightarrow ^{222}_{86}Rn + ^{4}_{2}\alpha$$

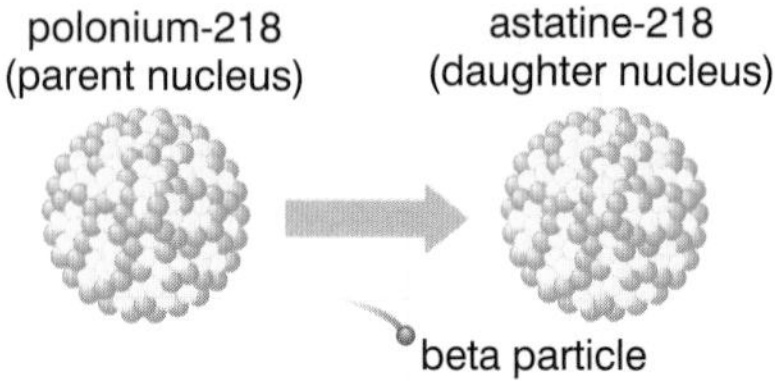

$$^{218}_{84}Po \rightarrow ^{218}_{85}At + ^{0}_{-1}\beta$$

CHECK YOURSELF QUESTIONS

Q1 Copy and complete this table to show the particles in these atoms.

Atom	Symbol	Number of protons	Number of neutrons	Number of electrons
Hydrogen	$^{1}_{1}H$			
Carbon	$^{12}_{6}C$			
Calcium	$^{40}_{20}Ca$			
Uranium	$^{238}_{92}U$			

Q2 The graph shows how the activity of a sample of sodium-24 changes with time. Activity is measured in becquerels (Bq).

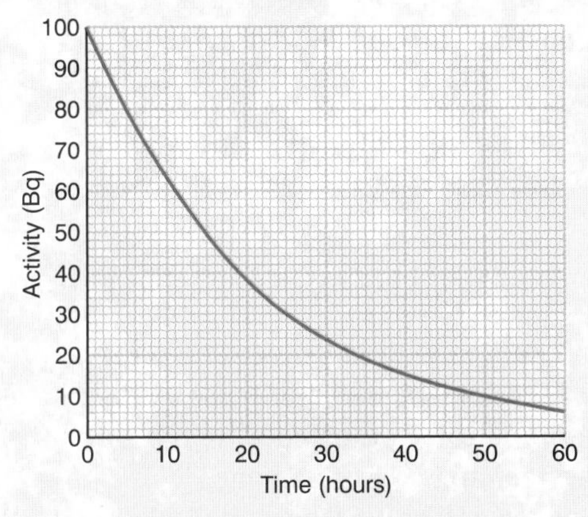

a Sodium-24 has an atomic number of 11 and a mass number of 24. What is the composition of the nucleus of a sodium-24 atom?

b Use the graph to work out the half-life of the sodium-24

Q3 The following equation shows what happens when a nucleus of sodium-24 decays.

$^{24}_{11}Na \rightarrow {}^{x}_{y}Mg + {}^{0}_{-1}\beta$

a What type of nuclear radiation is produced?

b What are the numerical values of x and y?

Answers are on page 149.

Uses and dangers of radioactivity

Radioactivity is dangerous

- Alpha, beta and gamma radiation can all damage living cells. Alpha particles, due to their strong ability to ionise other particles, are particularly dangerous to human tissue. Gamma radiation is dangerous because of its high penetrating power. Nevertheless, radiation can be very useful – it just needs to be used *safely*.

Uses of radioactivity

- Gamma rays can be used to kill bacteria. This is used in **sterilising** medical equipment and in preserving food. The food can be treated after it has been packaged.
- A **smoke alarm** includes a small radioactive source that emits alpha radiation. The radiation produces ions in the air which conduct a small electric current. If a smoke particle absorbs the alpha particles, it reduces the number of ions in the air, and the current drops. This sets off the alarm.

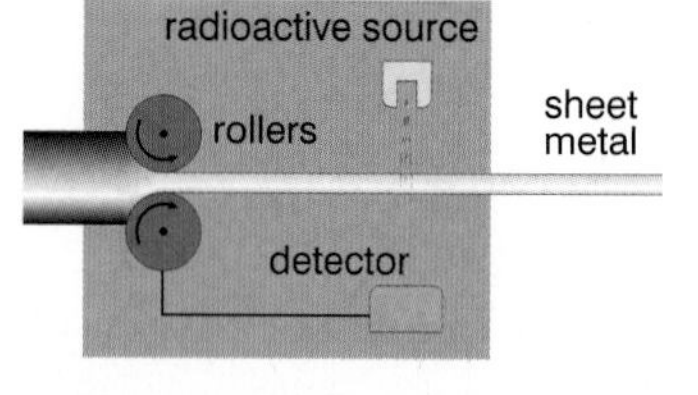

Sheet thickness control.

- Beta particles are used to monitor the **thickness** of paper or metal. The number of beta particles passing through the material is related to the thickness of the material.
- A gamma source is placed on one side of a **weld** and a photographic plate on the other side. Weaknesses in the weld will show up on the photographic plate.
- In **radiotherapy** high doses of radiation are fired at cancer cells to kill them.

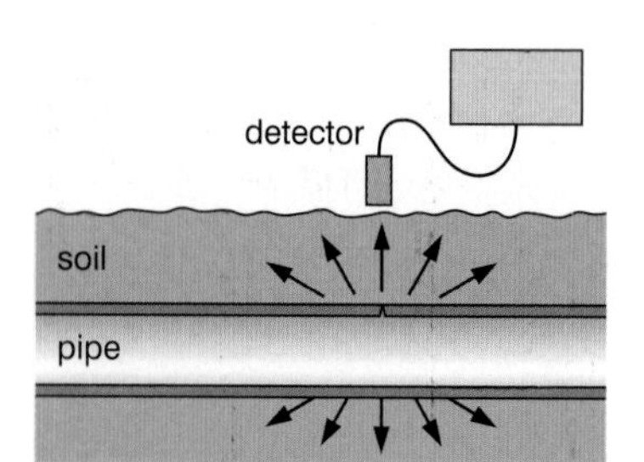

Tracers detect leaks.

- **Tracers** are radioactive substances with half-lives and radiation types that suit the job they are used for. The half-life must be long enough for the tracer to spread out and to be detected after use but not so long that it stays in the system and causes damage.
  - **Medical tracers** are used to detect blockages in vital organs. A gamma camera is used to monitor the passage of the tracer through the body.
  - **Agricultural tracers** monitor the flow of nutrients through a plant.
  - **Industrial tracers** can measure the flow of liquid and gases through pipes to identify leakages.

Using radiation to sterilise food has proved highly unpopular. People have the idea that the food would 'become radioactive'. How could a scientist demonstrate that this is not true? How could a scientist present the information to the public to help understanding? Is this a situation where what the public feels is more important than what a scientist would describe as 'the truth'? You should be able to give points on either side of the argument.

Radioactive dating

- Igneous rock contains small quantities of uranium-238 – a type of uranium that decays with a half-life of 4500 million years, eventually forming lead. The ratio of lead to uranium in a rock sample can be used to calculate the age of the rock. For example, a piece of rock with equal numbers of uranium and lead atoms in it must be 4500 million years old – but this would be unlikely as the Earth is only 4000 million years old!

- Carbon in living material contains a constant, small amount of the radioactive isotope **carbon-14**, which has a half-life of 5700 years. When the living material dies the carbon-14 atoms slowly decay. The ratio of carbon-14 atoms to the non-radioactive carbon-12 atoms can be used to calculate the age of the plant or animal material. This method is called **radioactive carbon dating**.

This woman's body was preserved in a bog in Denmark. Radioactive carbon-14 dating showed that she had been there for over 2000 years.

Nuclear fission

- In radioactive decay, each emission from the nucleus releases a small amount of energy. In nuclear fission, a neutron strikes a large nucleus. This splits the nucleus in two, forming **daughter nuclei**, releasing much *more* energy and two or three additional neutrons. If these neutrons split further nuclei, then a **chain reaction** is set up. A chain reaction can be used as an energy source for a power station.
- In a nuclear power station atoms of **uranium-235** are bombarded with neutrons and split into two smaller atoms (barium and krypton).
- The products of the fission are radioactive and so need to be disposed of very carefully.
- The amount of energy produced is considerable – a single kilogram of uranium-235 produces approximately the same amount of energy as 2 000 000 kg of coal!
- Do not confuse this process with the nuclear reactions taking place in the Sun. In the Sun, small atoms combine together to form larger atoms in a process known as **nuclear fusion**. This process also produces huge amounts of energy.

QUESTION SPOTTER

- It is very common for questions to ask *which type* of radiation is useful in a situation, and why. For example, for finding leaks in underground pipes a gamma source must be used because alpha and beta radiation would not be sufficiently penetrating to be detected at the surface.

CHECK YOURSELF QUESTIONS

Q1 What type of radiation is used in
a smoke detectors,
b thickness measurement,
c weld checking?

Q2 This question is about tracers.
a What is a tracer?
b The table below shows the half-life of some radioactive isotopes.

Using the information in the table only, state which one of the isotopes is most suitable to be used as a tracer in medicine. Give a reason for your choice.

Radioactive isotope	Half-life
lawrencium-257	8 seconds
sodium-24	15 hours
sulphur-35	87 days
carbon-14	5700 years

Q3 **a** When uranium-238 in a rock sample decays what element is eventually produced?

b Explain how the production of this new element enables the age of the rock sample to be determined.

Answers are on page 150.

Fundamental particles

Another layer down

- By the 1930s, the parts of the atom seemed to be clear – a central nucleus, consisting of protons and neutrons, with electrons orbiting around it.
- However, experiments involving particle accelerators and nuclear reactors have revealed many more particles. These provide evidence that protons and neutrons are themselves made from even smaller particles – **quarks**. These experiments also started to reveal particles of **anti-matter** – particles identical to ordinary particles, but with the opposite electrical charge. For example, the **positron** is identical to an electron, except that it has a **positive** charge.

IDEAS AND EVIDENCE

Experiments in particle physics require large amounts of energy and large particle accelerators. This makes research in this area very expensive. Most research is now a co-operative effort between different countries so the costs, and the benefits, are shared.

Tracks such as these provide evidence for other fundamental particles.

So what is fundamental?

- Fundamental particles are particles that have no internal structure and are not made from anything smaller – they are the 'end' of the story.
- All the evidence suggests that the electron is a fundamental, negatively-charged particle.
- Protons and neutrons are made of quarks. Two types of quark, called **up** and **down**, are needed. Three quarks combine to make a proton or a neutron.

Particle	Quarks
Proton	Up – up – down
Neutron	Up – down – down

- This helps explain the process of beta decay (see Revision Session 3). During the process of beta^- decay, a neutron is converted to a proton. Using the idea of quarks shows that at a deeper level, a down quark is being converted to an up quark. In beta^+ decay the reverse happens – an up quark is converted to a down quark, changing a neutron into a proton.

A* EXTRA

- Quarks cannot be detected directly. Evidence for their existence is gathered from their effect on other particles and the combinations of particles that are made during collisions in particle accelerators.

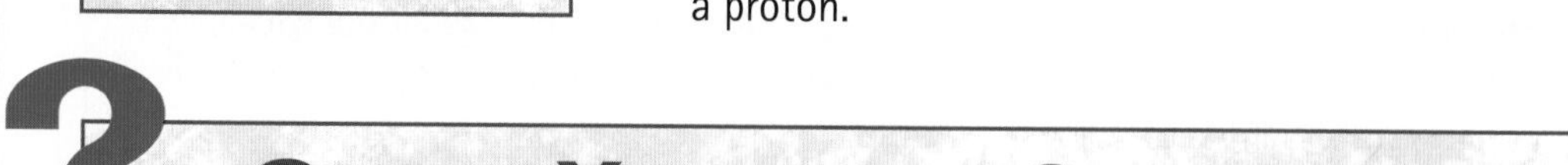

Check Yourself Questions

Q1 The up quark has an electric charge of $+^2/_3$ and the down quark has an electric charge of $-^1/_3$.

Show that these combine to give the correct charges for a proton (+1) and a neutron (0).

Q2 Particle research is very expensive. Suggest two reasons why countries are prepared to contribute to joint efforts in this area.

Answers are on page 150.

REVISION SESSION 6

Electron beams

Cathode rays

- Cathode rays were discovered in the late 1800s (see Revision Session 2). J.J. Thomson discovered that these rays consisted of a stream of **electrons** emitted from a heated cathode (a negative terminal).

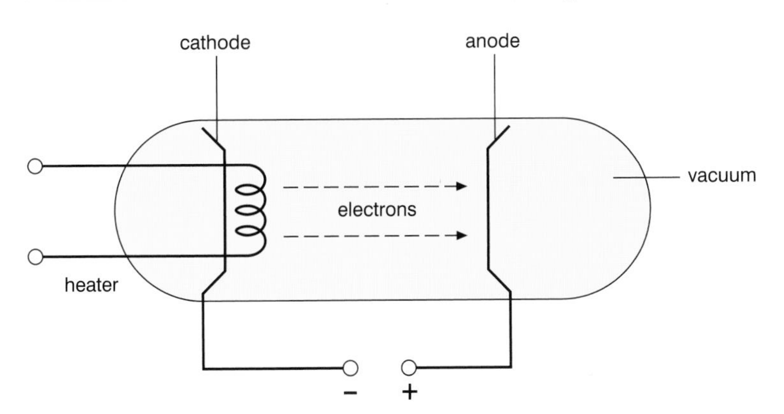

Thermionic emission – emitting electrons from a heated cathode

- The positive terminal (the anode) attracts the electrons from the cathode. The cathode is heated to increase the average energy of the electrons in the cathode, which makes it easier for the electrons to be 'pulled out'. The process of emitting electrons from a heated cathode is called **thermionic emission**.
- If there is a hole in the anode, a beam of electrons shoots through. The whole arrangement is then called an **electron gun**.

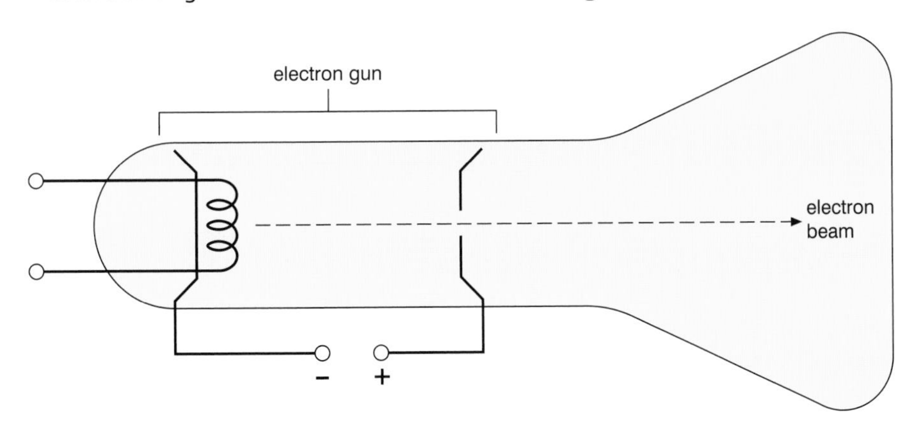

An electron gun.

A* EXTRA

- As the electron is accelerated between the cathode and anode, it gains kinetic energy. The energy gained is equal to $e \times V$, where e is the charge on the electron and V is the voltage between the terminals.

A* EXTRA

- The electron beam is an electric current. We can link the *number* of electrons moving per second to the *current* of the beam. Using the formula $I = \frac{Q}{t}$, if there is a current of 1.0 A, then there must be 1 coulomb of electric charge moving per second. 1 coulomb of electric charge is equivalent to 6.25×10^{18} electrons.

Deflecting the beam

- An electron beam is equivalent to an electric current, but without the wire! It can be deflected by **other electric charges** or by magnetic fields.

- In this diagram, the magnetic field is at right angles to the electron beam, so the beam is deflected. This is an example of the **motor effect** (see page 15).

Deflecting electrons with a magnetic field.

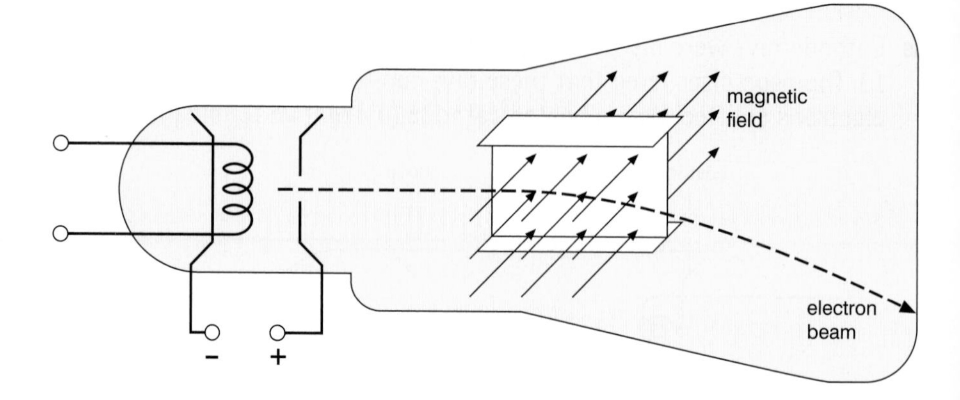

- In this diagram, the metal plates are charged, attracting the electron beam towards the positive plate and repelling it away from the negative plate.

Deflecting electrons with electric charges.

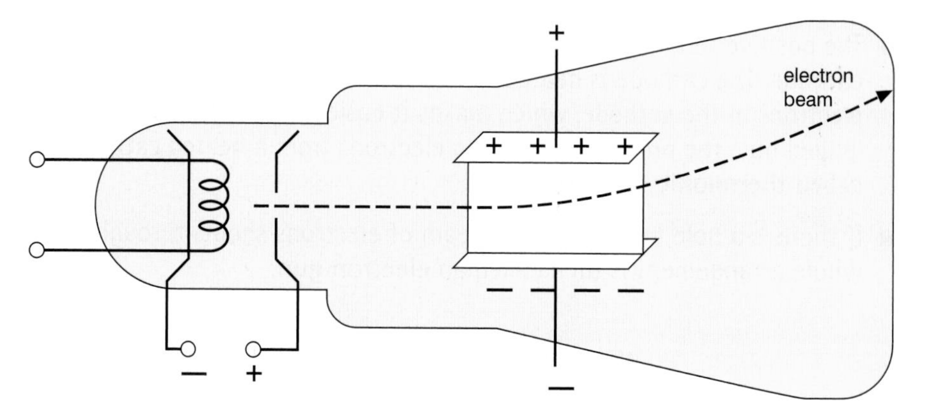

The cathode ray oscilloscope (CRO)

- In a CRO, the electron beam is directed towards a **fluorescent screen.** Where the beam hits the screen, the coating on the screen absorbs the energy from the electrons and releases the energy as light – a dot appears on the screen.

The cathode ray oscilloscope.

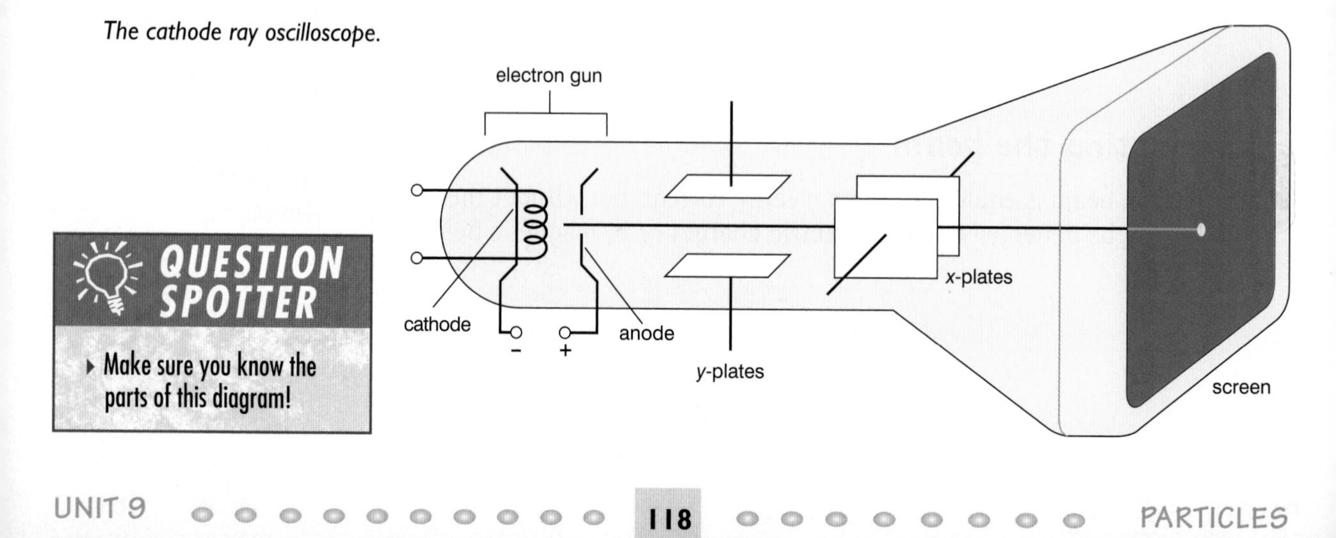

- Two sets of metal plates are used to deflect the beam – one pair in the **vertical** direction (called the **y-plates**) and one pair in the **horizontal** direction (called the **x-plates**). By controlling the voltages on these sets of plates, the dot can be moved to any position on the screen.

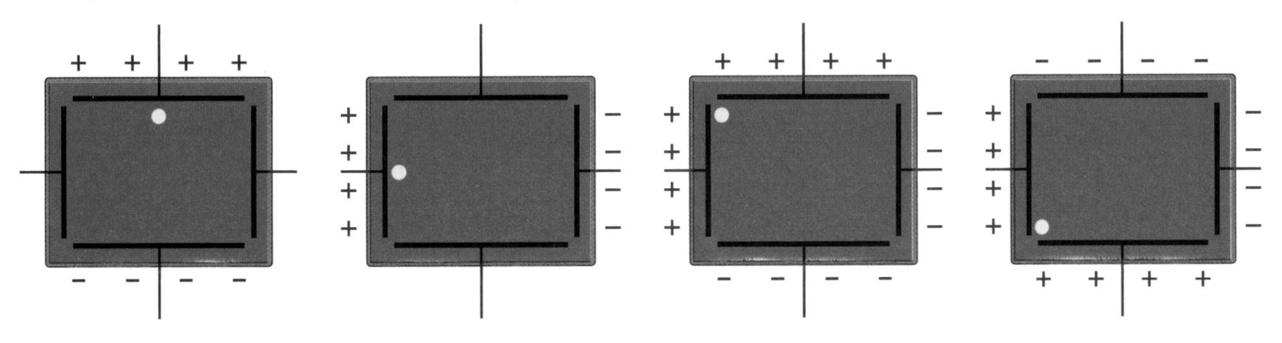

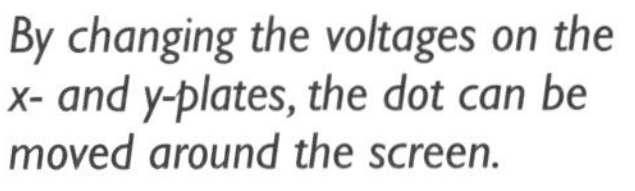

By changing the voltages on the x- and y-plates, the dot can be moved around the screen.

- A CRO has a circuit – the **time-base circuit** – that moves the dot across the screen, from left to right, at a constant speed and then returns the dot very quickly to the start. Repeating this process quickly means the dot appears as a **line** across the screen.

- A CRO can be used as a very high resistance voltmeter. If an unknown voltage is connected to the *y*-plates, the deflection produced in the beam can be compared to reference measurements and a value obtained.

- With the time-base circuit switched on and a variable signal connected to the *y*-plates, a CRO can be used to show how a waveform varies. This can be used to measure the frequency of a signal (see Unit 6 Revision Session 4).

Television tubes

- A cathode ray tube (CRT) used in a television is very similar to the CRO. For a colour television, three types of material are on the screen, divided into little dots, which provide the three primary colours of red, green and blue. Varying the intensity of the beam as it hits these dots makes other colours.

Cathode rays (electrons) in a television tube.

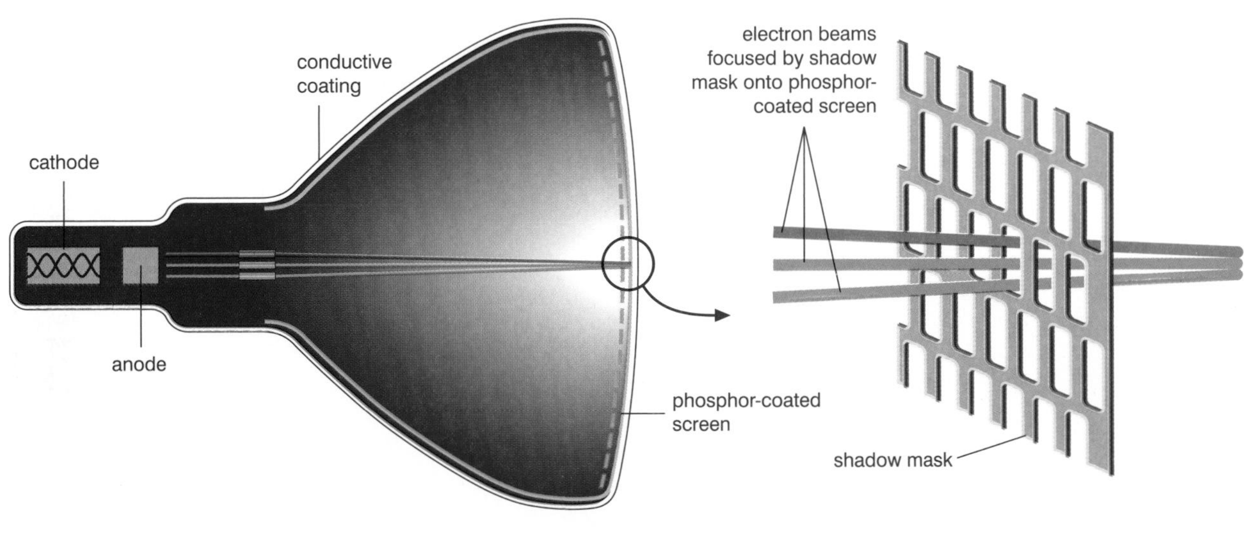

X-ray tubes

- To produce X-rays, the beam is directed at a crystal target. The energy from the electrons is absorbed by the target and re-emitted as X-rays.

Cathode rays (electrons) strike a target and some of their kinetic energy is converted to X-rays.

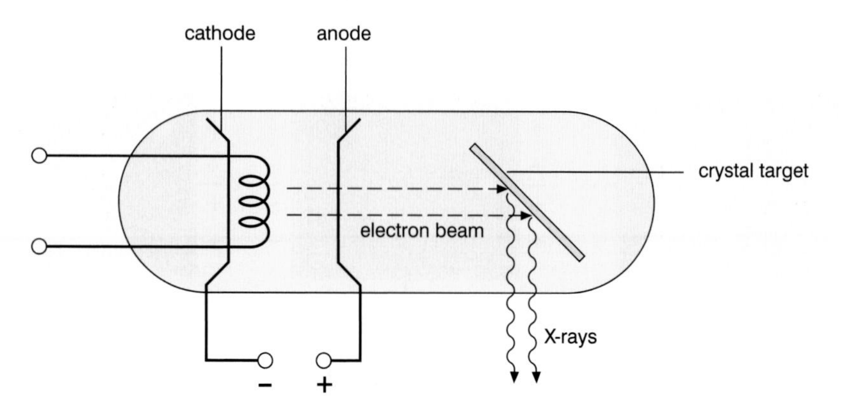

CHECK YOURSELF QUESTIONS

Q1 What is thermionic emission?

Q2 The diagram shows a CRO. Copy the diagram and add the labels in the correct places.

screen x-plates y-plates anode cathode electron gun

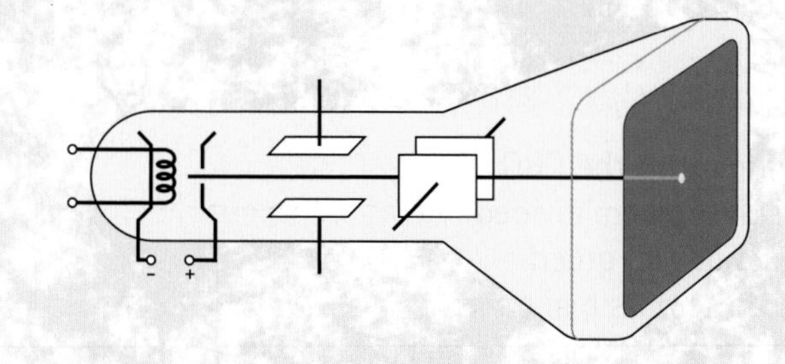

Q3 The diagram shows a trace on a CRO screen. The vertical setting is 0.2 V per division (square) and the timebase setting is 10 ms per division (square). 10 ms = 0.01 s.

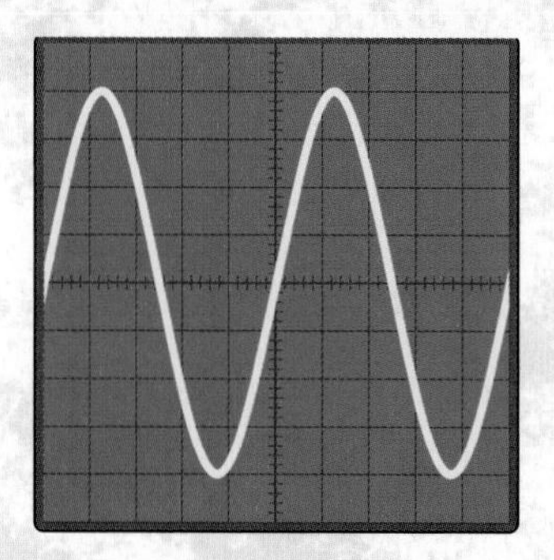

a What is the amplitude of the signal in volts?

b What is the time period of the signal (the time for one complete cycle)?

c What is the frequency of the signal?

Answers are on page 150.

UNIT 10: EXAM PRACTICE

Exam tips

- **Read each question carefully**; this includes looking in detail at any **diagrams**, **graphs** or **tables**.
  - Remember that any information you are given is there to help you.
  - Underline or circle the **key words** in the question and **make sure you answer the question that is being asked**.
- Make sure that you understand the meaning of the '**command words**' in the questions. For example:
  - '**Describe**' is used when you have to give the main feature(s) of, for example, a process or structure;
  - '**Explain**' is used when you have to give reasons, e.g. for some experimental results;
  - '**Suggest**' is used when there may be more than one possible answer, or when you will not have learnt the answer but have to use the knowledge you do have to come up with a sensible one;
  - '**Calculate**' means that you have to work out an answer in figures.
- Look at **the number of marks** allocated to each question and also the **space provided**.
  - Include at least as many points in your answer as there are marks. If you do need more space to answer, then use the nearest available space, e.g. at the bottom of the page, making sure you write down which question you are answering. **Beware of continually writing too much because it probably means you are not really answering the questions.**
- Don't spend so long on some questions that you don't have time to finish the paper.
  - You should spend approximately **one minute per mark**. If you are really stuck on a question, leave it, finish the rest of the paper and come back to it at the end.
- In short answer questions, or multiple-choice type questions, **don't write more than you are asked for**.
  - In some exams, examiners apply the rule that they only mark the first part of the answer written if there is too much. This means that the later part of the answer will not be looked at.
  - In other exams you would not gain any marks if you have written something incorrect in the later part of your answer, even if the first part of your answer is correct. This just shows that you have not really understood the question or are guessing.
- **In calculations always show your working**.
  - Even if your final answer is incorrect you may still gain some marks if part of your attempt is correct. If you just write down the final answer and it is incorrect, you will get no marks at all.
  - Also in calculations write your answer to as many **significant figures** as are used in the question.
  - You may also lose marks if you do not use the correct **units**.
- Aim to use **good English** and **scientific language** to make your answer as clear as possible.
  - In short answer questions, just one or two words may be enough, but in longer answers take particular care with capital letters, commas and full stops.
  - There should be an icon in the margin to warn you where there are separate marks for the quality of your English.
  - If it helps you to answer clearly, do not be afraid to also use **diagrams** in your answers.
- Some questions will be about scientific **ideas** and how scientists use **evidence**.
  - In these questions you may be given some information about an unfamiliar situation.
  - The answers to this type of question usually link to one of four areas: how scientists communicate their ideas; how scientific ideas can reflect the society in which the scientists work; how scientists can give different interpretations to the same evidence; how science can answer some questions but not others.
- When you have finished your exam, **check through** to make sure you have answered all the questions.
  - Cover your answers and read through the questions again and check your answers are as good as you can make them.

1 Anita wants to buy a car.

She looks at the information about the car.

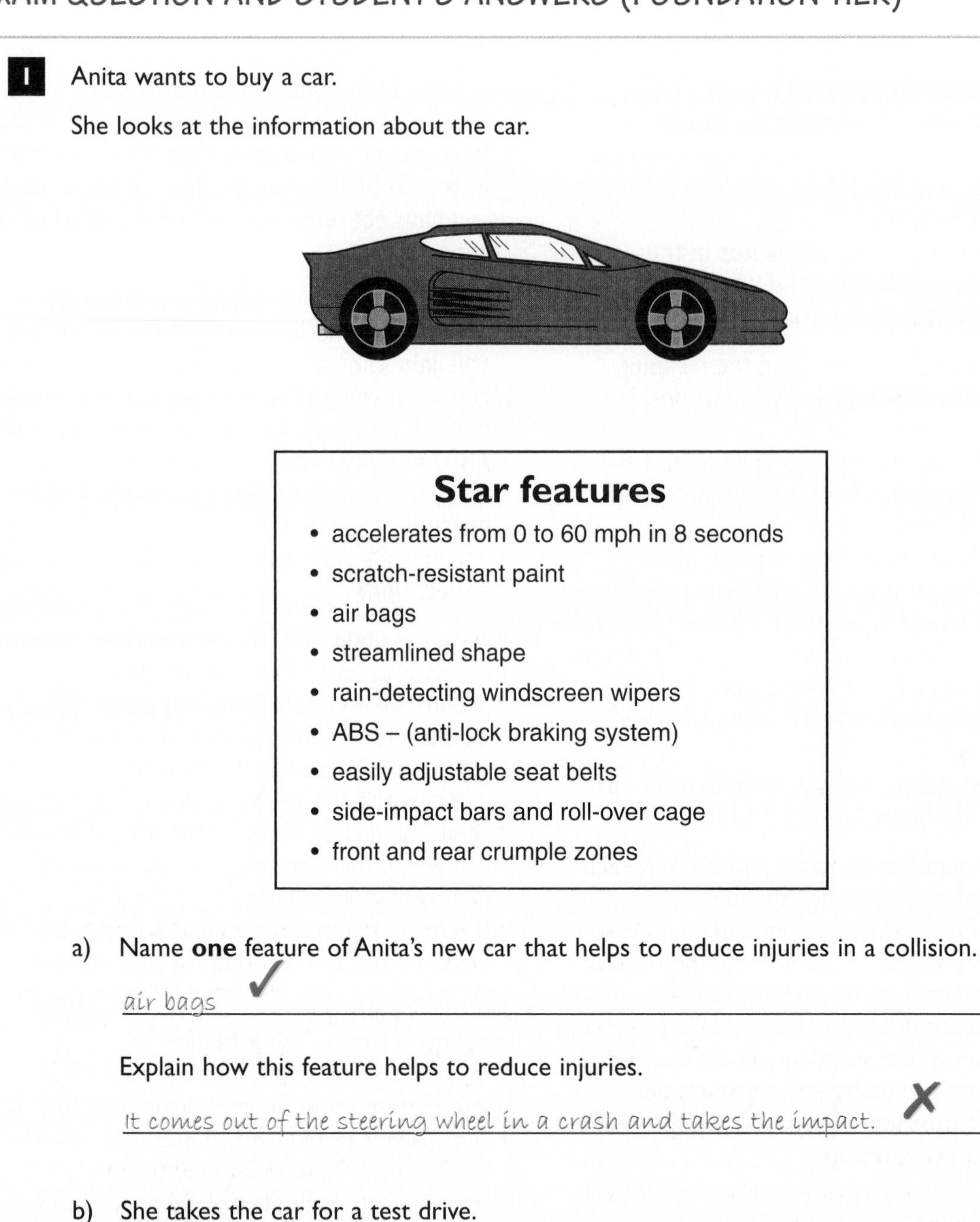

Star features

- accelerates from 0 to 60 mph in 8 seconds
- scratch-resistant paint
- air bags
- streamlined shape
- rain-detecting windscreen wipers
- ABS – (anti-lock braking system)
- easily adjustable seat belts
- side-impact bars and roll-over cage
- front and rear crumple zones

a) Name **one** feature of Anita's new car that helps to reduce injuries in a collision. (1)

air bags ✓

Explain how this feature helps to reduce injuries. (2)

It comes out of the steering wheel in a crash and takes the impact. ✗

b) She takes the car for a test drive.

She tests its stopping distance.

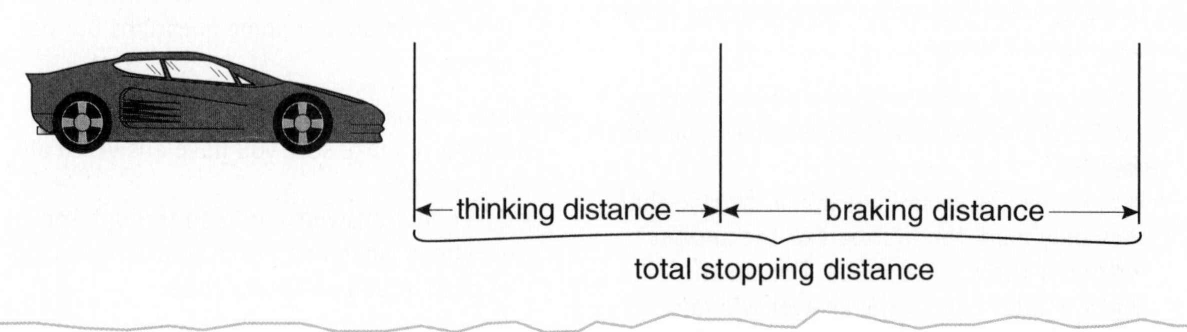

i) Which of the factors below would **increase** the total stopping distance?

Put a tick (✓) in each correct box.

dry road	☐
high speed	✓
new tyres	☐
rain	✓
slow speed	☐
worn, smooth tyres	✓

✓✓✓

(3)

ii) Write down **one** factor which will increase the **thinking** distance (1)

Drinking ✗

4/7

(Total 7 marks)

How to score full marks

a) One mark has been given for a correct choice. The other correct possibilities in the list are side-impact bars and roll-over cage, seat belts, front and rear crumple zones and ABS (anti-lock brakes). **The question asks for ONE feature – do not give more than one.**

No marks have been awarded for the explanation. 'It comes out of the steering wheel' says what the air bag **does**, not how it reduces injuries, so it scores no marks. 'Takes the impact' nearly scores a mark, but it does not have the key idea of **reducing** the impact. An answer for two marks here could have been 'The air bag reduces the impact (one mark) of the driver on the steering wheel (one mark)'. **The explanation needs to match the feature given above – go back and make another choice if you cannot give a clear explanation.** This question could also be asked at a C/D level, where a correct response would need to refer to reduced **acceleration** and hence reduced **forces**.

b) i) Three marks are awarded for correct choices. The question does not tell you how many boxes to tick, but there are three marks for this part of the question, so select three. **Always look at the number of marks available.**

ii) No mark has been awarded. Presumably the student means 'drinking alcohol' but it is not for the examiner to decide this. 'Drinking' could also refer to 'drinking water', which would not be a correct response. Correct responses here need to refer to **high speed** or some factor relating to the **driver**, for example being drunk, taking drugs (including prescription drugs), being tired or being distracted (possibly by the radio).

- A mark of 4 out of a possible 7 corresponds to a grade F on this Foundation Tier question.

QUESTION TO TRY (FOUNDATION TIER)

1 a) In the box are the names of five waves.

infra red **microwaves** **ultrasonic** **ultraviolet** **X-rays**

Which wave is used to:

i) send information to a satellite? (1)

ii) toast bread? (1)

ii) clean a valuable ring? (1)

b) The diagram shows four oscilloscope wave traces. The controls of the oscilloscope were the same for each wave trace.

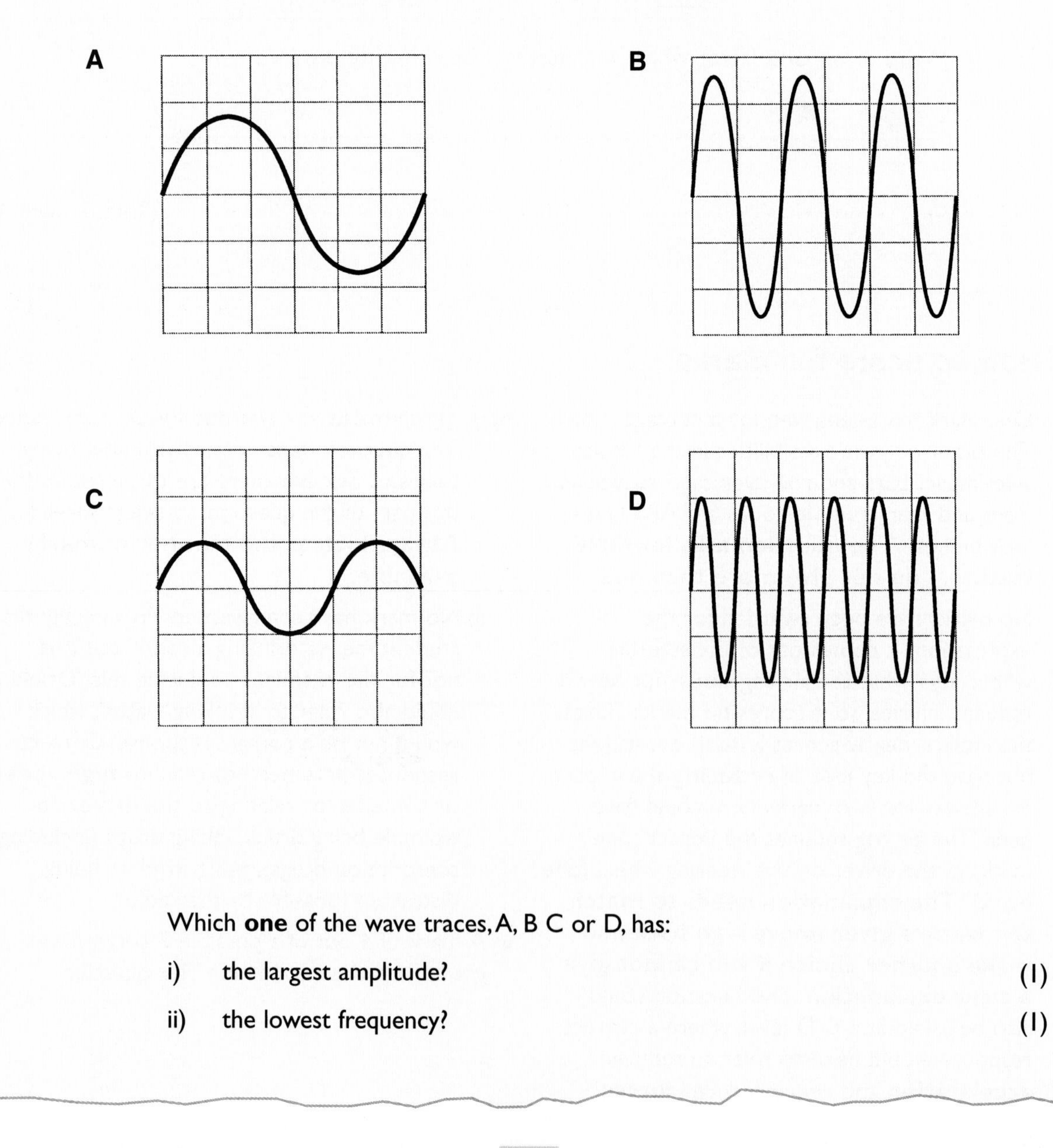

Which **one** of the wave traces, A, B C or D, has:

i) the largest amplitude? (1)

ii) the lowest frequency? (1)

c) The diagram shows a longitudinal wave in a stretched spring.

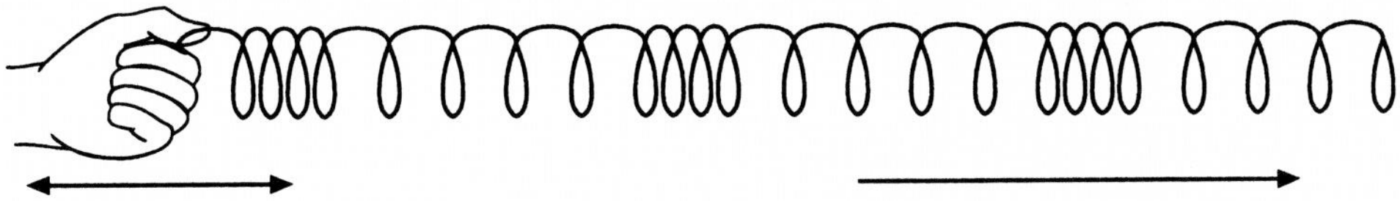

disturbance

direction of wave travel

Complete the sentence. You should put only **one** word in each space.

A longitudinal wave is one in which the .... causing the wave

is in the same .... as that in which the waves moves. (2)

d) Which **one** of the following types of wave is longitudinal? Draw a ring around your answer.

light wave **sound wave** **water wave** (1)

Answers are on page 151.

1 This question is about electrostatics.

a) There are two kinds electric charge.

Write down the names of both types of electric charge. (1)

positive and *negative* ✓

b) Leon wants to charge his plastic comb.

Write down one way he could charge his plastic comb. (2)

He could rub it. ✓ ✗

c) Leon holds his charged comb near some small pieces of paper.

Look at the diagram.

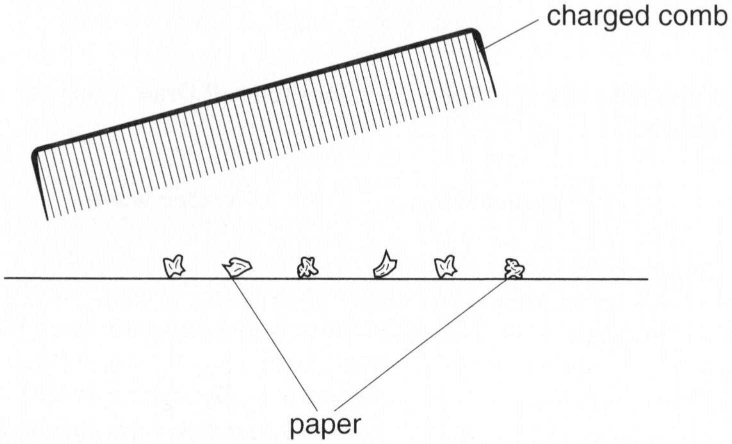

charged comb

paper

Suggest what might happen to the paper. (1)

They stick to the comb. ✓

d) Leon touches a metal radiator.

He gets an electric shock.

Describe how Leon gets an electric shock.

(One mark is for the correct use of scientific words.) (2+1)

The metal radiator is electric and gives Leon a shock. ✗ ✗ ✗

e) Leon paints cars.

Static electricity is useful in spraying paint.

i) Write down **one other** use of static electricity. (1)

A photocopier ✓

ii) Explain why static electricity is useful in spraying cars.

In your answer use ideas about electric charge.

(One mark is for linking ideas.) (3+1)

The paint is charged when it comes out of the sprayer. The car is also charged with the opposite charge. This makes the paint stick to the car much better. ✓✓

6/12

(Total 12 marks)

How to score full marks

a) The correct response has been given. The symbols '+' and '–' would have been acceptable.

b) 'He could rub it' scores one mark, although 'by friction' would have been a stronger phrase to use. There is a second mark for saying that the comb should be rubbed against **an insulator** (or you could give an example of an insulator, such as cloth). **Always check the number of marks available.**

c) One mark has been awarded for the correct response. The student indicates correctly that there will be an **attraction** between the comb and the pieces of paper.

d) This is a very vague answer. There are three marks available. The student can score any two. The correct response needs to realise the **Leon** has become **charged** (perhaps by friction against a carpet), that these charges **move** when he touches the radiator, **from Leon to the radiator**. The third mark is for using correct scientific words. Relevant words here are: charging, electrons, earth, earthing.

e) i) One mark has been awarded. Alternative correct responses would include inkjet printers, dust precipitators or crop spraying.

ii) Two marks have been awarded. The student seems to have an idea of what is happening, but has failed to use the correct scientific terms accurately. One mark has been awarded for the idea that opposite charges attract, but saying the paint 'sticks' to the car is not accurate enough to gain a second mark – the student needed to say that the paint is **attracted** to the car. In a similar way, saying the paint covers 'much better' is too vague. At this level, the student should refer to the paint being attracted to **the whole object**, even parts not in a direct line, or that **less paint is wasted**. Another approach would be to state that **like charges repel** (one mark) and so the paint forms **a fine spray** (one mark), which produces **an even coat** (one mark). The student does, however, gain the 'linking ideas' mark for linking the idea of opposite charges repelling to a consequence of this.

- A mark of 6 out of a possible 12 corresponds to a grade D on this Foundation/Higher Tier overlap question.

1 This question is about using scientific ideas and evidence.

The Octon Electricity Company want to build a new power station.

It will burn fossil fuels.

Look at the map.

It shows the area around Octon where they want to build the power station.

There are two available sites, **A** and **B**.

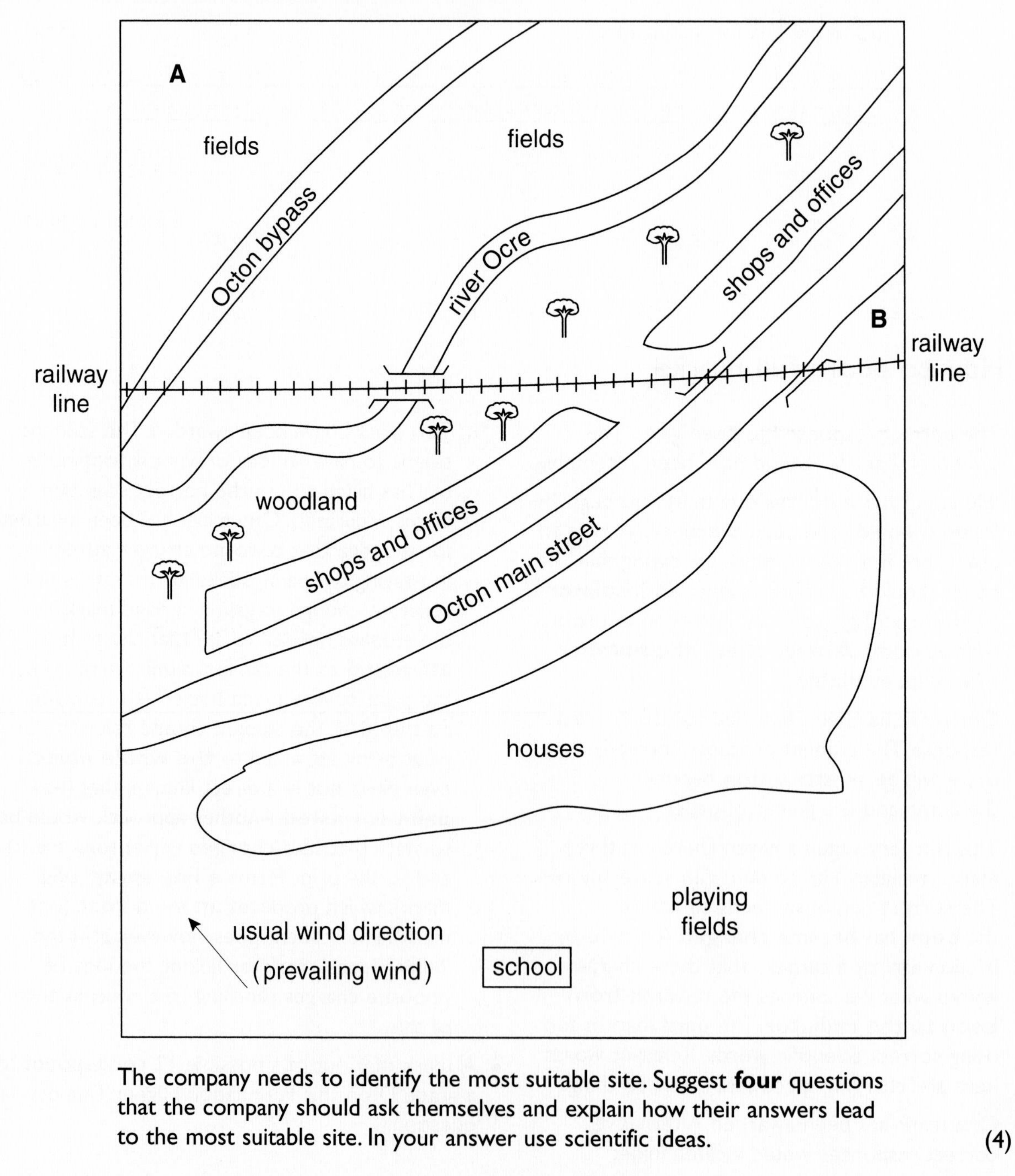

The company needs to identify the most suitable site. Suggest **four** questions that the company should ask themselves and explain how their answers lead to the most suitable site. In your answer use scientific ideas. **(4)**

2 The diagram shows the 'plum pudding' model of an atom.

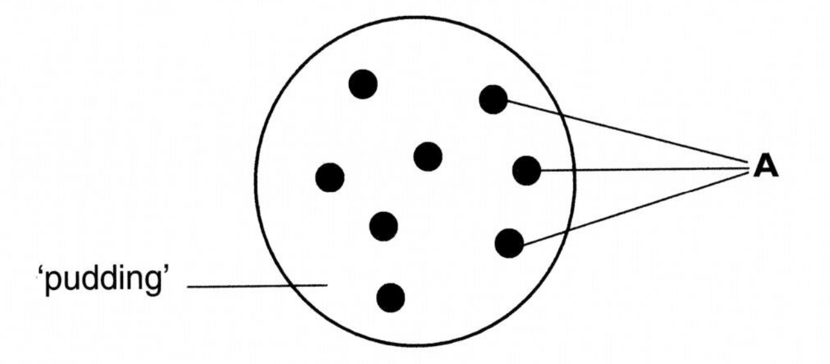

a) Scientists thought that the 'pudding' was positively charged.

i) Name the particles labelled **A** in the diagram. (1)

ii) Complete this sentence by choosing the correct words from the box.

negatively charged	**positively charged**	**uncharged**

The particles labelled **A** are.. (1)

b) A new model of an atom was suggested by Rutherford and Marsden.

They fired alpha particles at thin metal foil.

Alpha particles are positively charged.

In their model each atom had a nucleus.

The diagram below shows the path of an alpha particle as it passes the nucleus of an atom.

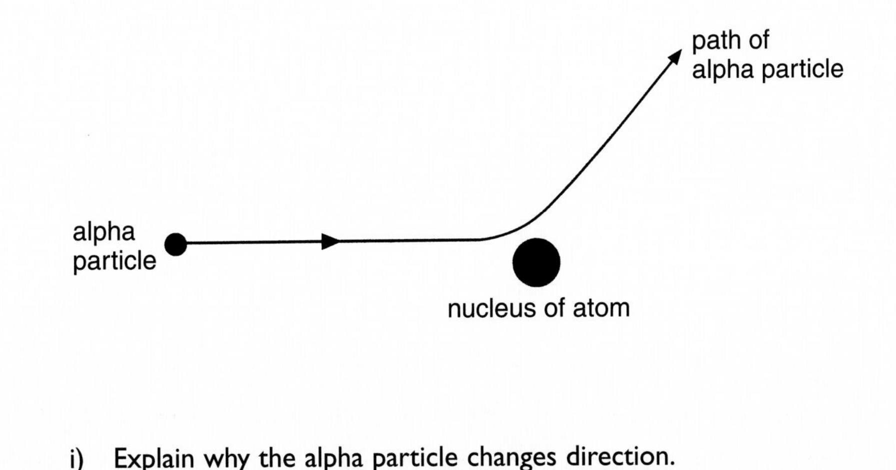

i) Explain why the alpha particle changes direction. (2)

ii) The diagram shows different paths taken by alpha particles when they were fired by Rutherford and Marsden at the thin metal foil.

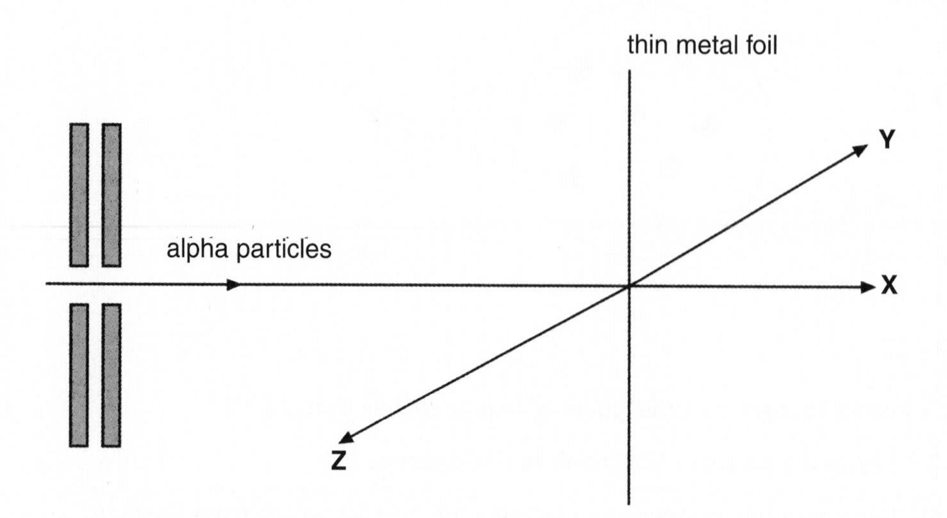

The boxes on the left show some observations from the experiment.

The boxes on the right give their explanations.

Copy the diagram and draw a straight line from each observation to its explanation.

One has been done for you. (2)

Observation	Explanation
some alpha particles travel along path **Y**	because the nucleus has a very large positive charge and a large mass
most of the particles follow path **X**	because the nucleus is very small
some alpha particles rebound backwards along path **Z**	because most of each atom is empty space
very few alpha particles follow path **Z**	because the nucleus is positively charged

Answers are on page 151.

1 a) There are three main ways by which radio waves arrive at the receiver from the transmitter.

One of these is by **ground waves**. Name the other two. (2)

1 sky waves ✓

2 surface waves ✗

b) Ground waves do not travel in a straight line. They follow the curvature of the Earth.

Name the effect that causes them to travel in this way. (1)

diffraction ✓

c) A radio station in La Coruna, Spain broadcasts MW radio waves of wavelength 639 m.

These waves travel through the air at 300×10^6 m/s.

Calculate the frequency of these waves. (4)

speed = frequency x wavelength ✓

300×10^6 = frequency x 639 ✓

frequency = $\frac{300 \times 10^6}{639}$ = 469 000 Hz ✓ ✓

d) The ionosphere is part of the Earth's upper atmosphere. It is affected by the Sun. Radio waves from La Coruna are absorbed by the ionosphere during the day and reflected by it at night. Scientists wanted to use this during an eclipse to help them to study the Sun.

Read this passage published by 'The Radiocommunications Agency' at the time of the total solar eclipse in the UK in 1999.

> Scientists from Oxfordshire's Rutherford Appleton Laboratory are asking the public to help with some unique experiments during the total eclipse on August 11th this year. At this time, RAL, in partnership with *The Daily Telegraph* and BBC Online are asking for your help in answering the question 'Is our Sun getting brighter?'
>
> You can help by simply listening to your radio.
>
> One station, La Coruna in northern Spain, broadcasts on 639 MW. In the UK, this station can only be heard at night. If we hear it on the morning of August 11th, we will know the eclipse has had a dramatic effect on the ionosphere.

i) The last time a total solar eclipse was seen in the UK was in 1927. Why was this experiment not done then? (1)

The Sun didn't have an effect then. ✗

ii) Why do you think the Rutherford Appleton Laboratory asked *The Daily Telegraph* and BBC Online to help? (1)

So more people would know about it. ✓

iii) *The Daily Telegraph* and BBC Online did not publish the full text of the Rutherford Appleton Laboratory's research project. Suggest a reason why not. (1)

It would take up too much space. ✗

7/10 (Total 10 marks)

How to score full marks

a) One mark has been awarded for 'sky waves'. 'Space waves' should be given for the second mark – this is basic recall.

b) 'Diffraction' is correct – it is the spreading out of waves around corners.

c) One mark has been given for the correct equation. **Always give the equation you are using.** The values have been substituted correctly, so one mark has been awarded. **Always write out the equation with the values in before you actually do the calculation.** The answer has been worked out correctly, earning one mark. The unit has been given, also earning one mark. **Always remember the unit.** Note that 469 kHz would also gain the marks. Also notice that the answer has been **rounded to 3 significant figures** – do not write out every digit the calculator shows you.

d) This section relates to **ideas and evidence in science.** The key is to **use clues in the information given** to give **reasonable suggestions.**

i) No mark has been awarded – this is **not** an example of a reasonable suggestion. Although the Sun **may** have changed over the last century, it is **more likely** that the reason lies with developments here on Earth. Responses worthy of credit would include the idea that fewer households had radios or that 1927 was in the early days of research into the ionosphere when less was known about it.

ii) One mark has been awarded. One of the **key** aspects of ideas and evidence in science is **how** and **why** scientists get their ideas across to different audiences. In this case, **using clues from the passage**, it is clear that the scientists wanted as many people as possible to be involved, which for modern times means using the broadcast media (radio, TV) and newspapers. They would do this **so that they could gather as much data as possible**, which is an idea that would also score a mark.

iii) No mark has been awarded. The key idea here is that the press release would be **written for a specific audience**, in this case the general public. It is likely that the full report would contain too much technical language for the general public to follow, which might well put the public off helping.

- A mark of 7 out of a possible 10 corresponds to a low grade B on this Higher Tier question.

1 Bobby goes on holiday in a caravan.

He uses a wind turbine to make electricity.

a) The turbine turns the simple a.c. generator shown below.

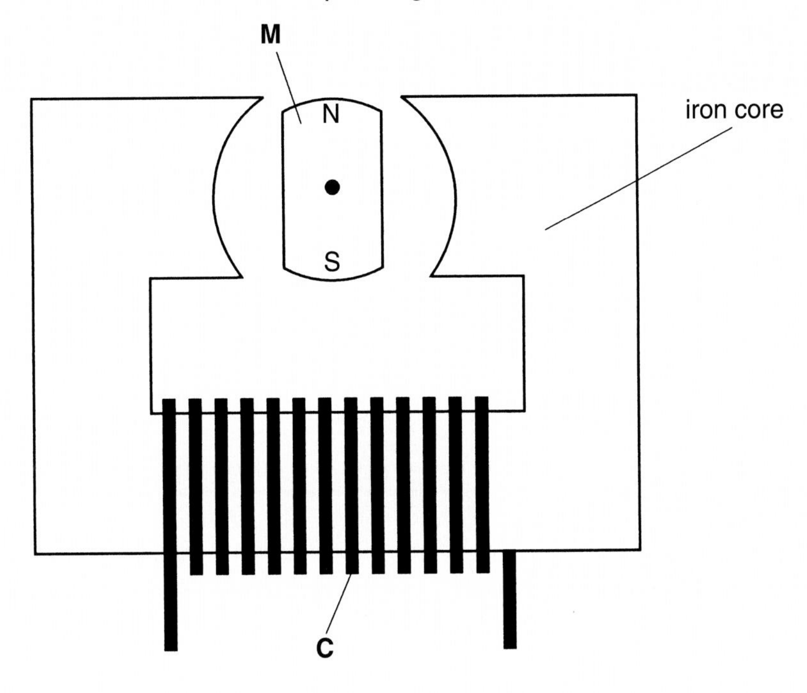

i) The term a.c. stands for **alternating current**.
Explain the difference between alternating current and direct current. (1)

ii) Explain how the simple generator shown above generates alternating current.
Include the names of the parts labelled **M** and **C**. (4)

b) The generator is connected to the caravan by long cables.
Bobby is worried about energy losses in the cable.

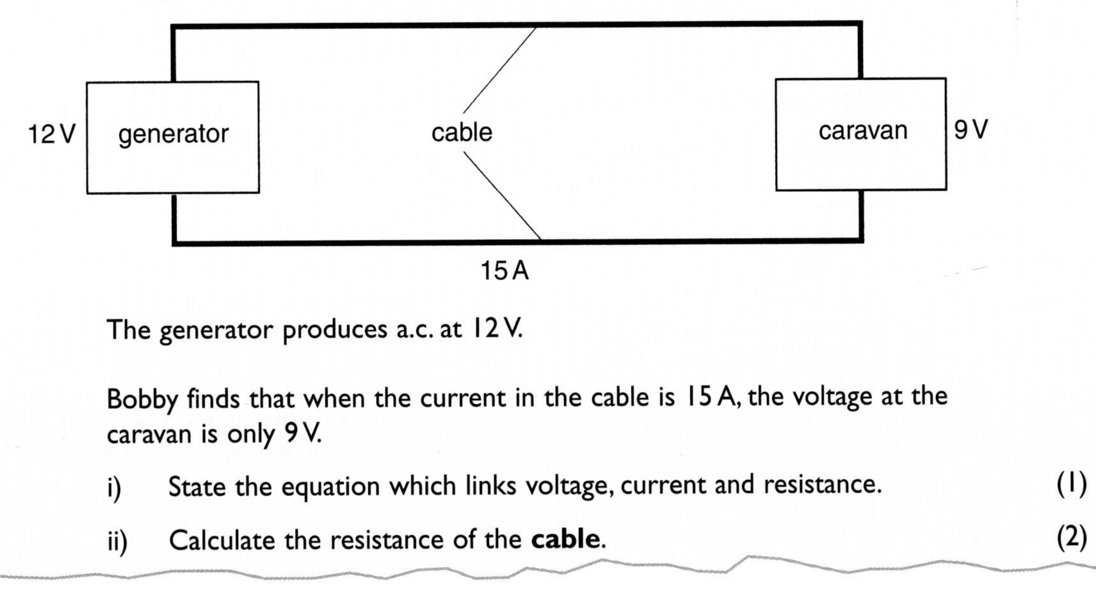

The generator produces a.c. at 12 V.

Bobby finds that when the current in the cable is 15 A, the voltage at the caravan is only 9 V.

i) State the equation which links voltage, current and resistance. (1)

ii) Calculate the resistance of the **cable**. (2)

2 Our Sun is a star.

Eventually it will become a white dwarf.

Write about the life cycle of our Sun.

In your answer you should write about:

- its formation
- the processes that happen during its life. (4)

3 A hot air balloon is tied to the ground by two ropes.

The diagram shows the forces acting on the balloon.

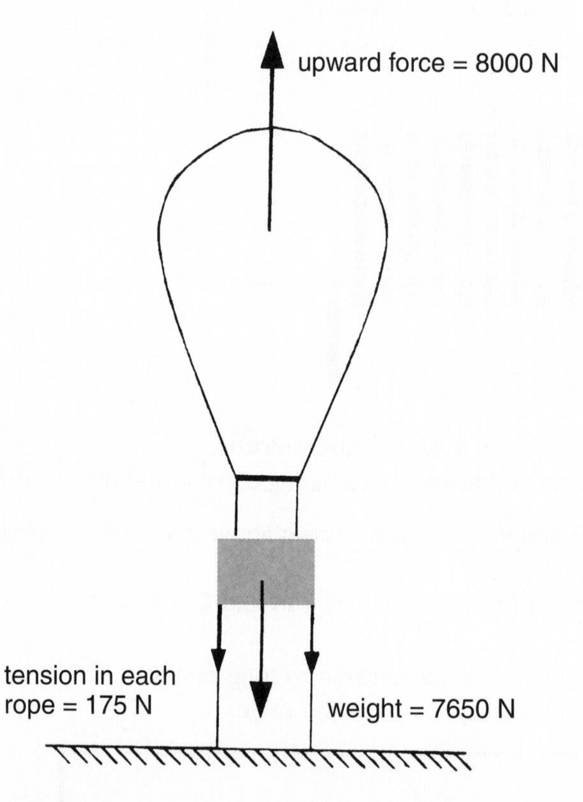

The ropes are untied and the balloon starts to move upwards.

a) Calculate the size of the unbalanced force acting on the balloon. State the direction of this force. (2)

b) The mass of the balloon is 765 kg. Calculate the initial acceleration of the balloon. (3)

c) Explain how the acceleration of the balloon changes during the first ten seconds of its flight. (4)

d) When the balloon is still accelerating, the balloonist throws some bags of sand over the side. Explain how this affects the acceleration of the balloon. (2)

Answers are on page 152.

ANSWERS

UNIT 1: ELECTRICITY

1 Electric circuits (page 5)

Q1 a A_1 reading 0.2 A, A_2 reading 0.2 A.

Comment The current is always the same at any point in a series circuit.

b A_4 reading 0.30 A, A_5 reading 0.15 A.

Comment The ammeter A_6 has been placed on one branch of the parallel circuit. Ammeter A_5 is on the other branch. As the lamps are identical the current flowing through them must be the same. Ammeter A_4 gives the current before it 'splits' in half as it flows through the two parallel branches.

Q2 Reading on V_1 = 6 V.

Comment The potential difference across the battery is 9 V. This must equal the total p.d. in the circuit. Assuming there is no loss along the copper wiring the p.d. across the lamp must be 9 − 3 = 6 V.

Q3 a 0.33 A

Comment Use the equation $I = Q/t$, $I = 10/30 = 0.33$ A.

b (i) 10 C; (ii) 36 000 C.

Comment Rearranging the formula, $Q = It$,
(i) $Q = 10 \times 1 = 10$ C, (ii) $Q = 10 \times 60 \times 60 = 36\,000$ C.

2 What affects resistance? (page 9)

Q1 a

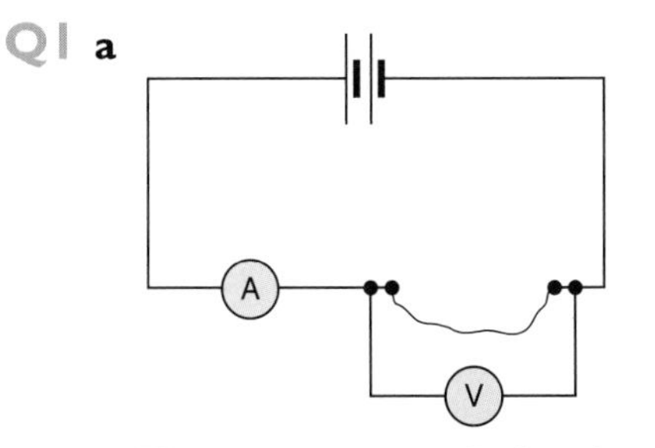

The resistance is calculated using Ohm's law, $R = V/I$.

Comment Remember that the ammeter must be in series with the nichrome wire, but the voltmeter must be in parallel.

b (i) The resistance would double.
(ii) The resistance would be halved.

Comment Resistance is directly proportional to length. Resistance is inversely proportional to cross-sectional area. If necessary look back at the models on page 7.

Q2 a 24 V

Comment $V = IR = 2 \times 12 = 24$ V

b 20 V

Comment $V = IR = 0.1 \times 200 = 20$ V

c 0.12 A

Comment $I = V/R = 12/100 = 0.12$ A

d 23 A

Comment $I = V/R = 230/10 = 23$ A

e 60 Ω

Comment $R = V/I = 6/0.1 = 60\ \Omega$

f 23 Ω

Comment $R = V/I = 230/10 = 23\ \Omega$

Q3 a The resistance increases as the voltage increases.

Comment If the resistance were constant the line would be straight with a constant gradient. It curves in such a way that increasing voltage produces a smaller increase in the current, so resistance must be increasing.

b As the voltage increases the temperature of the wire must also be increasing.

Comment This is a common characteristic of non-ohmic resistors – that is, resistors that do not obey Ohm's law.

3 Power in electrical circuits (page 12)

Q1 920 W

Comment Power = $V \times I$ = 230 × 4 = 920 W. Don't forget the unit. Power is measured in watts.

Q2 **a** 24p

Comment Cost = units × 8 = 1 × 3 × 8 = 24p. Remember that a unit is a kilowatt-hour so the 1000 W has to be converted into kilowatts, that is 1 kW.

b 8p

Comment 200 W = 0.2 kW. Cost = units × 8 = 0.2 × 5 × 8 = 8p.

Q3 1800 kJ

Comment Energy = power × time (s) = 3000 × 10 × 60 = 1 800 000 J or 1800 kJ. Remember that 1 watt is 1 joule/second.

4 Static electricity (page 14)

Q1 **a** Electrons are rubbed off the surface of the plastic onto the cloth.

Comment Static electricity is produced by removing electrons from one insulator to another. An excess of electrons leads to a negative charge.

b The positively charged rod induces a negative charge on the surface of the paper near to the rod. Opposite charges then attract.

Comment Electrostatic induction is a common phenomenon. Remember this is how you can make a balloon stick to the ceiling.

Q2 **a** The passenger is charged by friction as her clothes rub against the seat covers. The car is also charged by friction with the road and the air. Touching metal allows the charge to flow to earth.

Comment Remember static electricity is produced by friction.

b Touching a door handle after walking on a synthetic carpet, or removing clothing.

Comment Synthetic fibres are more likely to cause this effect than natural fibres. This must be due to the different atom arrangements.

Q3 Lightning, fuelling aircraft, in grain silos.

Comment Static electricity can cause explosions in any situations when there are fuels in the gaseous form or where there are very fine particles of solid material (dust).

Unit 2: Electromagnetic effects

1 Electromagnetism (page 17)

Q1 **a** No. Aluminium is not magnetic.

Comment Remember that very few metals are magnetic (e.g. iron, cobalt and nickel). The electromagnet would separate steel cans as steel is an alloy containing iron.

b Switch off the current to the electromagnet.

Comment The magnetic field is only present when the current is flowing. An electromagnet is a temporary magnet.

Q2 Increasing the current flowing through the coil. Increasing the number of coils. Adding a soft iron core inside the cardboard tube.

Comment These will produce a stronger magnetic field.

Q3 When the switch is pressed, the electromagnet attracts the hammer support and the hammer hits the bell. The movement of the hammer support breaks the circuit and so the electromagnet ceases to operate. The hammer support then returns to its original position, forming the circuit again and the process is repeated.

Comment The electromagnet is constantly activated and then deactivated. This means that the hammer will continually hit the bell and then retract.

2 Electromagnetic induction (page 21)

Q1 If the rotation of the armature is producing electricity which is being withdrawn via A and B, then it is a generator. If, alternatively, electricity is being supplied via A and B to produce rotation then it is a motor.

Comment The same device can be used as either a motor or a generator. A motor needs an electrical supply in order to produce movement. In a generator the movement is used to generate electricity.

Q2 **a** Iron.

Comment The core must be a magnetic material. It concentrates the magnetic effect.

b A magnetic field is produced.

Comment Remember that a current flowing in a wire will produce a magnetic field around it.

c A current is induced.

Comment The varying magnetic field in the core induces a current in the secondary coil.

d 24 V.

Comment $V_p/V_s = N_p/N_s$; $V_p/14 = 12/7$; $V_p = 12/7 \times 14 = 24$ V.

Q3 An electric current produces a heating effect in a wire, thus transferring energy as heat. To reduce energy loss, electricity is transmitted at as low a current as possible, which means transmitting at very high voltage.

Comment The heating effect of an electric current is considered in Unit 1.

3 More electromagnetic devices (page 24)

Q1 Changing the volume control:
- reduces a resistance, which
- increases the current, so
- the magnetic field of the coil becomes stronger, so
- the attraction/repulsion with the magnet becomes stronger, and so
- the paper cone in the speaker vibrates with a bigger amplitude, which
- sets the air vibrating more strongly, so
- David hears a louder sound.

Comment This is quite a long list. Judge how long your answer should be by the number of marks available.

Q2 When Helen records the magnetic pattern on the tape it does not store perfectly. Each further copy 'scrambles' the magnetic pattern slightly more, particularly if there are additional signals (noise) in the wires that might be caused by other parts of the recording equipment.

Comment Remember that the tape stores information as a magnetic pattern – it does not store the music itself!

Q3 The information is stored as a magnetic pattern on the tape. Placing the tape near a permanent magnet may re-organise the pattern on the tape, destroying the information that was wanted.

Comment This question could also relate to computer floppy disks, which also store information as magnetic patterns.

Unit 3: Electronics and Control

1 Logic gates (page 28)

Q1 AND gate.

> **Comment** For the safe to open, both input conditions must be met.

Q2 a

Inputs			Output
A	B	C	
0	0	0	1
0	1	1	0
1	0	1	0
1	1	1	0

> **Comment** 'Way point' C is the output from an OR gate. This becomes the input for the NOT gate.

b

Inputs			Output
A	B	C	
0	0	0	1
0	1	0	1
1	0	0	1
1	1	1	0

> **Comment** 'Way point' C is the output from an AND gate. This becomes the input for the NOT gate.

Q3

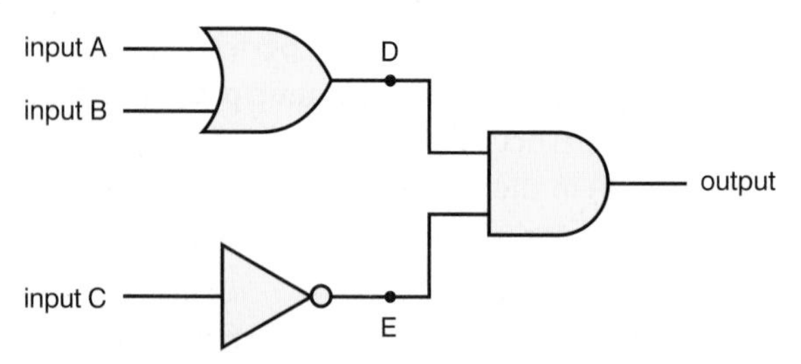

Inputs			D	E	Output
A	B	C			
0	0	0	0	1	0
0	0	1	0	0	0
0	1	0	1	1	1
0	1	1	1	0	0
1	0	0	1	1	1
1	0	1	1	0	0
1	1	0	1	1	1
1	1	1	1	0	0

> **Comment** Add 'way points' D and E.
> D is A OR B.
> E is NOT C.
> Output is D AND E.

2 Input sensors and output devices (page 32)

Q1 The input voltage to the logic gate will start low and gradually change to high. This is because when it is cold, the thermistor has a high resistance, so it takes a large share of the supply voltage. As the temperature rises, the resistance of the thermistor falls, so it takes a smaller and smaller share of the voltage.

> **Comment** Look again at the description given in the 'Using a light-dependent resistor (LDR)' section in the text. The heat sensor works in a similar way, with the thermistor replacing the LDR.

Q2 8 V

> **Comment** $V = (1000/1500) \times 12$

Q3
- Logic gate cannot provide enough current to run heater.
- It isolates logic gate from higher power heater circuit (preventing damage to logic gate).

> **Comment** Be clear what you are referring to – say 'the logic circuit cannot provide enough current', not just 'it cannot provide enough current'.

3 Electronic systems (page 36)

Q1 a Input: pressure sensor and light sensor
Processor: logic gate
Output: buzzer

b AND gate

> **Comment** Pardeep wants the buzzer to sound if *both* sensors send a high signal.

c Remove the AND gate and replace it with a NOR gate latch (see Revision session 1)

> **Comment** If the pressure sensor is the 'set' sensor for the latch, the alarm will still ring when the 'burglar' steps off the sensor. The master switch is then the 'reset' for the latch and can turn off the buzzer.

d

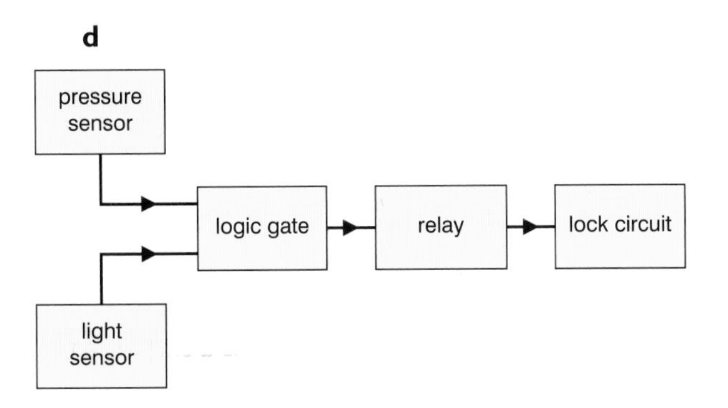

Comment Remember that a relay is always needed to connect logic circuits to higher current (higher power) circuits.

Q2

Inputs			Outputs		
A	B	C	R	Y	G
0	0	0	1	0	0
0	0	1	1	0	0
0	1	0	1	0	0
0	1	1	1	1	0
1	0	0	0	0	1
1	0	1	0	0	1
1	1	0	0	0	1
1	1	1	0	1	0

Comment Remember to include 'way points' after each logic gate – take it a step at a time!

UNIT 4: FORCES AND MOTION

1 The effects of forces (page 41)

Q1 A – the wooden block would not move; B – the car would accelerate; C – the ice skater would maintain a constant speed.

Comment A – The forces are balanced and so the stationary wooden block would not move. B – The resultant force would be 4000 N. This unbalanced force will cause the car to accelerate. (As the car moves faster the resistive force caused by air resistance will increase and eventually reach 6000 N). C – The forces are balanced. As the skier is moving she will continue to do so at the same speed.

Q2 a

Comment Friction will always oppose motion and so the arrow points in the opposite direction to the direction of motion.

b The frictional forces slow the skier down.

Comment As the skier travels downhill the friction force opposes the motion.

c Using narrower skis, or using wax on the skis.

Comment Lubricant reduces frictional forces.

Q3 a *Stage 1* – The skydiver is accelerating. The downward force of gravity is greater than the upward force caused by air resistance. *Stage 2* – The sky diver is travelling at constant speed. The forces of gravity and air resistance are balanced. *Stage 3* – The sky diver is slowing down. The force caused by the air in the parachute is greater than the force of gravity. *Stage 4* – The sky diver is travelling at a constant speed. The forces are balanced again.

Comment In questions of this sort the first thing to do is to decide which forces will be acting on the object. The next thing is to decide whether the forces are balanced or unbalanced. If they are unbalanced the skydiver will be either accelerating or decelerating. If they are balanced the skydiver will either be travelling at constant speed or not moving. In stage 2 the skydiver will have reached the terminal speed and because the forces are balanced will not accelerate or decelerate.

b The force of gravity is balanced by the upward force of the ground on the sky diver.

Comment The upward force from the ground is equal to the skydiver's weight.

Q4 a The graph is a straight line from the origin to the (12 cm, 6.0 N point). After that it becomes increasingly curved and less steep.

Comment Don't be fooled into drawing a single 'best fit' line – there are two distinct parts to this graph.

b Elastic up to the (12 cm, 6.0 N) point, plastic after that.

Comment The straight-line part is elastic, the curved part is plastic.

c 0.5 N/cm.

Comment Work out the gradient of the graph in the straight-line (elastic) section.

2 Velocity and acceleration (page 48)

Q1 a The speed remained constant.

Comment If the line on a distance–time graph has a constant gradient then the speed is constant.

b 600 m.

Comment This can be read off the graph: after 20 s the car had travelled 200 m, after 60 s it had travelled 800 m. So the distance travelled is 800 – 200 = 600 m.

c 15 m/s.

Comment Speed = distance/time, v = 600/40 = 15 m/s. Don't forget the units.

d It stopped.

Comment The line is horizontal during this time, indicating that the car was not moving (no distance was travelled).

Q2 a 2.0 m/s/s.

Comment Acceleration = change in velocity/time, a = (1.5 – 0)/0.75 = 2 m/s/s.

b 2.25 m.

Comment Total distance = area under the line = $(1/2 \times 0.75 \times 1.5) + (1/2 \times 2.25 \times 1.5)$ = 2.25 m. *Note:* the tractor showed constant acceleration from A to B and then constant deceleration from B to C.

Q3 a 2.5 m/s/s.

Comment Acceleration = change in speed/time, a = (30 – 0)/12 = 2.5 m/s/s. Don't forget the units!

b 2500 N.

Comment Force = mass × acceleration, F = 1000 × 2.5 = 2500 N.

Q4 a 120 m northwards and 90 m westwards.

Comment speed = distance/time, so distance = speed × time

b 150 m at an angle 37° to north, a bearing of 323°.

Comment Draw the diagram to scale and measure the length and angle.

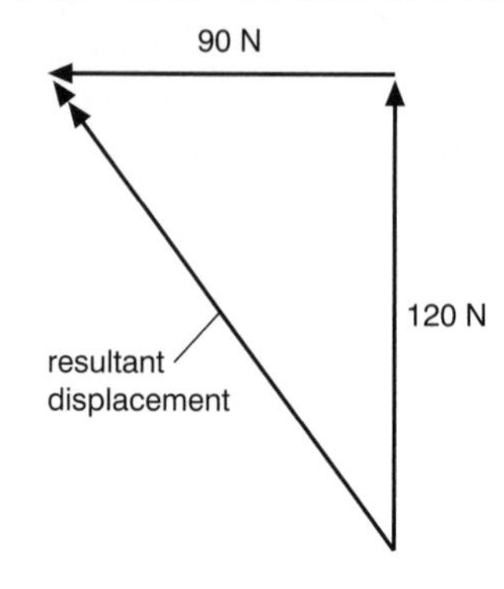

3 Vehicle safety features (page 50)

Q1 The seat belt stretches a little to allow the passenger to slow down slowly. This reduces the acceleration and the forces involved.

Comment This type of question could be linked to work on stretching materials, see Revision session 1.

Q2 **a** An air bag takes a short time to inflate. In a very high speed crash, the passenger may have already hit a hard surface before the air bag has inflated.

b Allowing the air bag to deflate gives the passenger even more time to stop, reducing the acceleration and the forces even more. If the air bag stayed fully inflated, it would hold the passenger in their seat too strongly, reducing the stopping time and so increasing the acceleration.

Comment Again, the key idea is to spread the stopping over a longer time – this reduces the acceleration, which reduces the forces, which reduces the injuries. Learn this sequence!

4 Using the equations of motion (page 53)

Q1 1440 m.

Comment Use $s = ut + \frac{1}{2}at^2$
Remember that u = 0 m/s (the train 'sets off from a station) and that the time needs to be in seconds (2 minutes = 120 seconds).

Q2 18 m/s^2.

Comment Use v = u + at and re-arrange.
Remember u = 0 m/s ('from rest').

Q3 –11.25 m/s^2.

Comment Use $v^2 = u^2 + 2as$ and re-arrange.
Remember that the minus sign in the acceleration indicates that the car is slowing down.

5 Motion in two dimensions (page 55)

Q1 **a** The path starts curving down as soon as he releases the ball, so it will miss the coconut (unless he throws it very fast!).

Comment Make sure the path starts to curve down *immediately after its release.* Many candidates draw the ball travelling horizontally for a while and then dipping downwards – this is incorrect. Gravity pulls the path down as soon as the ball is released.

b Aim slightly above the coconut so that the ball hits it on the way down.

Comment Don't fall into the trap of saying 'throw it upwards' – that would cause it to land on Danny's head!

Q2 **a** Both land at the same time.

Comment Gravity causes both balls to fall. The sideways motion does not affect gravity, so both balls land at the same time.

b 4 s.

Comment Horizontally, the ball travels at constant speed (ignoring air resistance) so use speed = distance/time and re-arrange to find the time taken.

6 Momentum (page 59)

Q1 **a** 7500 kg m/s.

Comment momentum before = 500 kg × 15 m/s = 7500 kg m/s,
momentum after = 500 kg × 30 m/s = 15 000 kg m/s.

b 1500 N.

Comment net force = change in momentum/time taken for change = 7500/5

c The engine will need to provide force to overcome friction as well as providing force to accelerate.

Comment It is very common to confuse forces and energy – make sure your answer refers to forces. Engines provide additional *force* to overcome *forces* acting against them; engines provide additional *energy* to compensate for *energy losses*, e.g. through heating.

Q2 a 25 000 kg m/s; –15 000 kg m/s.

Comment momentum = mass $\times$ velocity, taking 'to the right' as a positive direction

b 10 000 kg m/s

Comment (25 000) + (-15 000). Do not forget to take account of the directions.

c 10 000 kg m/s.

Comment Use the principle of conservation of momentum.

d 5.7 m/s to the right.

Comment velocity = momentum/mass = 10 000/1750, positive value indicates 'to the right'.

Q3 Before he steps off, momentum = 0 (since it is stationary). If David gains momentum to the right, then the skateboard must gain momentum to the left, for the total momentum to remain zero.

Comment Remember that momentum has a direction, so the overall momentum can be zero even though both objects are moving.

7 Turning forces (page 62)

Q1 800 Nm.

Comment Moment = force $\times$ distance

Q2 C.

Comment C has the widest base and is heavier at the bottom (more glass).

Q3 2 m to the right of the pivot (on Jane's side).

Comment At the moment, Jane's clockwise moment is 300 N $\times$ 4 m = 1200 Nm. Freddy's anticlockwise moment is 450 N $\times$ 4 m = 1800 Nm. Freddy needs to provide a moment of 600 Nm clockwise to balance the see-saw. Freddy weighs 300 N, so he needs to sit 2 m from the pivot (300 N $\times$ 2 m = 600 Nm).

Unit 5: Energy

1 Where does our energy come from? (page 64)

Q1 a A source that cannot be regenerated – it takes millions of years to form.

Comment A common mistake is to say that it is a source that 'cannot be used again'. Many energy sources cannot be used again but they can be regenerated (e.g. wood).

b Coal, oil and natural gas.

Comment These are the fossil fuels. Substances obtained from fossil fuels such as petrol and diesel are not strictly speaking fossil fuels.

c Coal.

Comment Coal is becoming increasingly more difficult to mine as more inaccessible coal seams are tackled. The 300 year estimate could be very optimistic.

Q2 a The total energy used increased from 1955 to 1975 and then levelled out.

Comment Always describe a graphical trend carefully. Be as precise as you can.

b The amount of coal used has significantly reduced. The amounts of oil, gas and nuclear fuel used have significantly increased.

Comment It is important to refer to each of the energy sources given on the graph.

Q3 a Any two from: strength of the wind, high ground, constant supply of wind, open ground.

Comment Higher ground tends to be more windy than lower ground. However, the wind farm cannot be built too far away from the centres of population otherwise costs will be incurred in connecting to the National Grid.

b *Advantage:* renewable energy source, no air pollution. *Disadvantage:* unsightly, takes up too much space.

Comment Wind turbines can be very efficient. However, they need to be reasonably small and so a large number are needed to generate significant amounts of electricity. Environmentally, although they produce no air pollution they do take up a lot of land.

2 Transferring energy (page 68)

Q1 The thin layers will trap air between them. Air is an insulator and so will reduce thermal transfer from the body.

Comment Remember that air is a very good insulator. Trapping it between layers of clothing means that convection is inhibited as well as conduction.

Q2 The molecules that are able to leave the surface of the liquid are those that are moving the fastest, that is, the molecules with the most energy. The average energy of the remaining molecules is therefore reduced and so the liquid is cooler.

Comment The temperature of a liquid is related to the average speed of its molecules. If the temperature is reduced, on average, the molecules will move more slowly.

Q3 Energy from the hot water is transferred from the inner to the outer surface of the metal by conduction. Energy is transferred through the still air by radiation.

Comment In a question like this it is important to take the energy transfer stage by stage. Thermal transfer through a solid involves conduction. Thermal transfer through still air must involve radiation. In fact, convection would be occurring and it is very likely that the air behind the radiator would be moving (convention currents).

3 Specific heat capacity (page 70)

Q1 a 45 600 J.

Comment $E = m \times c \times \theta$ Remember to use the temperature *change*, 80 °C.

b 105 600 J

Comment $E = m \times c \times \theta$ Remember to use the temperature *change*, 80 °C.

c Copper

Comment To achieve the same temperature rise, much more energy needs to be transferred to the aluminium. For cooking, this is a waste of energy that would be better going into the food.

Q2 Water has a high specific heat capacity. This means that it can remove large amounts of energy from the hot engine without boiling itself.

Comment Formula 1 cars do not have water cooling; they rely on airflow to remove heat. A Formula 1 car overheats very quickly if the car is not moving.

Q3 a 1 600 000 J (1.6 MJ).

Comment $E = m \times c \times \theta$ Remember to use the temperature *change*, 40 °C.

b 1 600 000 J

Comment Energy is conserved, so energy taken in at night = energy released during the day.

c If copper was heated to the same temperature, it would not store so much energy – the lower specific heat capacity shows that less energy transfer is needed to achieve the same temperature rise.

Comment Alternatively, if the copper was heated until it stored the same energy, the temperature would be much higher, which would be undesirable (and dangerous) in a domestic heating system.

4 Work, power and energy (page 73)

Q1 12.5 m.

Comment Work done = $F s$
$F = 400 \times 10 = 4000$ N
$s = W/F = 50\,000/4000 = 12.5$ m

Q2 **a** 6000 J.

Comment Energy transferred = $30 \times 200 = 6000$ J

b 33.3 W.

Comment Power = work done/time taken = 6000/180 = 33.3 W. Remember that a watt is 1 joule/s so the time must be in seconds.

Q3 **a** 10 500 J.

Comment P.E. = $m g h = 35 \times 10 \times 30 = 10\,500$ J

b 24.5 m/s.

Comment Assuming all the potential energy is transferred to kinetic energy:
K.E. = $1/2\, mv^2$, so $v^2 = 2$ K.E.$/m$
$= 2 \times 10\,500/35 = 600$, and $v = 24.5$ m/s

c Some of the gravitational potential energy will have been transferred to thermal energy due to the friction between the sledge and the snow.

Comment Energy must be conserved but friction is a very common cause of energy being wasted, that is, being transferred into less useful forms.

Unit 6: Describing waves

1 The properties of waves (page 77)

Q1 **a** D.

Comment The crest is the top of the wave.

b A and E.

Comment The wavelength is the length of the repeating pattern.

c B.

Comment The amplitude is half the distance between the crest and the trough. If you had difficulty with parts a to c look back at the diagram on page 74.

d 512.

Comment Frequency is measured in hertz and 1 Hz = 1 cycle (or wave) per second.

Q2 0.33 m.

Comment $v = f \times \lambda$ or $\lambda = v/f$
$\lambda = 3 \times 10^8 / 9 \times 10^8 = 0.33$ m

Q3 1.33 m

Comment $v = f \times \lambda$ or $\lambda = v/f$
$\lambda = 340/256 = 1.33$ m

2 The electromagnetic spectrum (page 79)

Q1 **a** Light.

Comment The sensitive cells form a part of the eye called the retina.

b Ultraviolet.

Comment Suntan lotions contain chemicals that absorb some of the UV radiation from the Sun before it can act on the skin.

c Microwaves.

Comment The key word here was 'rapid'. Electric cookers make use of infrared radiation for cooking but microwaves produce much more rapid cooking.

d X-rays.

Comment A gamma camera would not be as good for this purpose as the gamma rays penetrate the bone as well as the flesh.

e Infrared.

Comment Remote car locking sometimes uses infrared beams.

Q2 a They can pass through soft tissue and kill cancer cells.

Comment Gamma rays are useful because of their great penetrating power. This is also their disadvantage. Consequently they have to be used extremely carefully. They are the highest energy waves in the electromagnetic spectrum.

b They can damage healthy cells and cause cancer because of their very high energy.

Comment As you might have guessed this is an important point and one that is often tested in exams!

c They have different frequencies and wavelengths.

Comment All waves in the electromagnetic spectrum have different frequencies and wavelengths.

d 300 000 000 m/s.

Comment All waves in the electromagnetic spectrum travel at the same speed in a vacuum – the speed of light.

3 Light reflection and refraction (page 84)

Q1 a In reflection light changes direction when it bounces off a surface. In refraction the light changes direction when it passes from one medium to another.

Comment In reflection the angle of incidence and the angle of reflection are the same. In refraction the angle of incidence does not equal the angle of refraction, as the ray of light bends towards or away from the normal.

b (i)

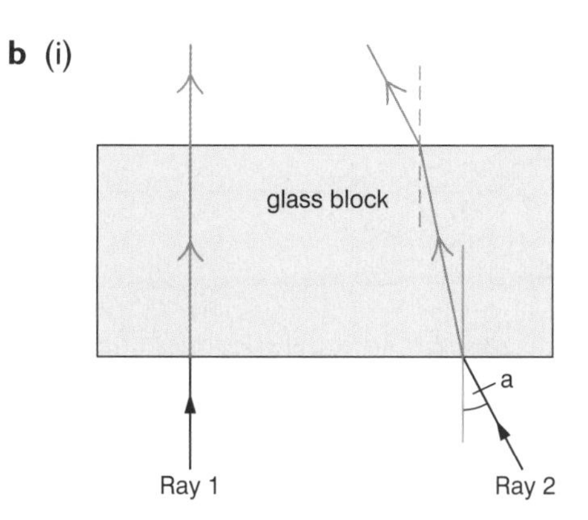

b (ii) The angle of incidence.

b (iii) The normal line.

Comment The ray of light will only be bent if it hits the block at an angle. In both cases the speed of the light in the block will be slower than in air.

Q2 a

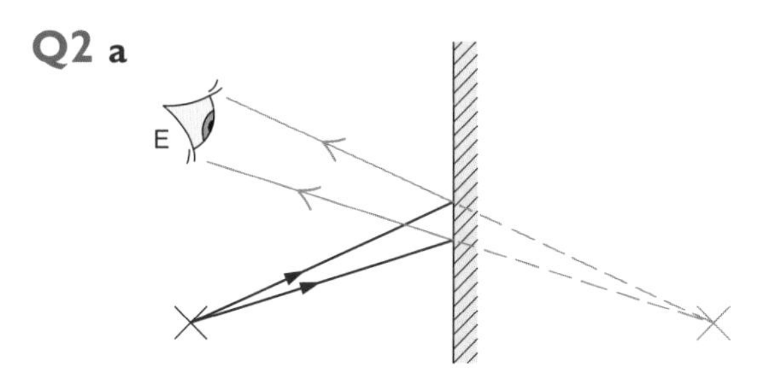

Comment Remember that with a plane mirror the image is the same size as the object and will be the same distance behind the mirror as the object is in front. When drawing the ray diagram check that the angles of incidence and reflection are the same.

b Light will be reflected off the hair onto the first mirror. If the angle is correct the reflected ray will then travel to the second mirror. If the angle is correct it will then be reflected into the eye.

Comment Don't forget that the light travels from the object to the eye! In an examination you might find it easier to draw a diagram showing the passage of the ray of light. In this way you can show the correct angles on the mirrors.

Q3 a

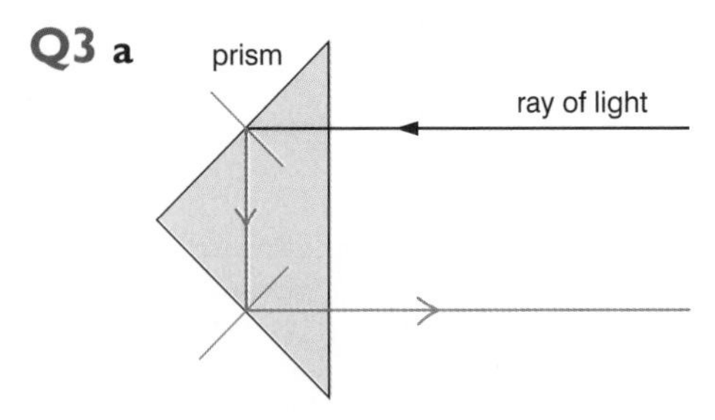

Comment You should draw a normal line to ensure that for each reflection the angles of incidence and reflection are the same.

b Binoculars, bicycle reflector, cat's eyes.

Comment Periscopes are also often made using prisms rather than mirrors.

c R, T, T.

Comment Total internal reflection occurs only when the angle of incidence is equal to or greater than the critical angle.

4 Sound waves, resonance and musical instruments (page 88)

Q1 a (i) Vibrations. (ii) Compressions and rarefactions travel through the air.

Comment These are both common questions, so don't forget. Sound is a longitudinal wave. The compressions and rarefactions are small differences in air pressure.

b Sound waves cannot travel through a vacuum. Radio waves don't require a medium.

Comment This is a big difference between sound waves and electromagnetic waves.

c Particles in water are closer together so the vibrations are passed on more quickly.

Comment The denser the material the greater the speed of sound. In concrete, for example, the speed of sound is 5000 m/s.

d A wine glass will vibrate with a particular natural frequency. If the opera singer sings a note of exactly the same frequency, it may cause resonance in the glass. This means the glass will vibrate with a much larger amplitude, which could break the glass.

Comment Remember that during resonant effects, there is a transfer of energy from the driving system into the resonating system.

Q2 a Above the limit of audible sound.

Comment The frequency of ultrasonic waves is too high for the human ear to hear.

b The motion of the medium is parallel to the direction of movement of the wave.

Comment This contrasts with transverse waves where the medium moves at right angles to the movement of the wave.

c The wavelength decreases, the frequency remains the same.

Comment You should remember that $v = f \times \lambda$; v decreases, f stays the same so λ must decrease.

Q3 375 m.

Comment $s = v \times t \times 1/2 = 1500 \times 0.5 \times 0.5 = 375$ m

5 Interference (page 90)

Q1 Choose any of the examples from the table on page 89.

Comment In a 'describe an experiment' question, think of four things:
- What equipment is needed?
- How is it used?
- What is observed?
- How does this relate to the effect being investigated?

Q2 The noise from the engine and the upside down signal produce destructive interference all the time, so the pilot hears no noise.

Comment This could also be applied to jamming a radio signal.

Q3 Different theories about a problem often lead to different predictions about new situations. Experimental work is needed to check these predictions. Experimental work can also suggest new lines of enquiry for theoretical work.

Comment There are several areas within physics where this question could be asked. The answer is always valid.

Unit 7: Waves for communication

1 Information transfer and storage (page 93)

Q1 **a** Turning the torch on and off.

b Holly's eye.

c Holly's brain.

Comment Not all communication systems have to be high-tech! Just consider carefully the steps involved in transferring the information.

Q2 The record produces an analogue signal. Any scratch on the disc will add noise to the signal. The record player will amplify this noise and it will be heard through the speakers. A CD produces a digital signal. A scratch will also add noise to this signal, but the processor is able to reconstruct the signal properly since it only needs to tell the difference between high and low parts of the signal.

Comment Learn which devices use analogue signals and which use digital. Be careful with cassette tape and video tape – both types of signal can be used.

2 Radio systems (page 96)

Q1

Radio station	Frequency	Wavelength
BBC Radio 2	89.9 MHz	3.3 m
BBC Radio 4	1.5 MHz	198 m
BBC Radio 5 Live	330 kHz	909 m

Comment Remember the wave equation, $v = f\lambda$. Rearrange the formula to find $f = v/\lambda$ and $\lambda = v/f$. 1 MHz = 1 000 000 Hz. 1 kHz = 1000 Hz.

Q2 A repeater amplifies the signal that arrives. This makes up for attenuation effects. A regenerator reconstructs the original signal. The effect on the transmitted signal is that a regenerator will re-transmit the original, whereas a repeater will transmit a signal with amplified noise on it.

Comment A digital signal is needed for the regenerator to work.

Q3 **a** TV and radio signals have different wavelengths. As Reece lives in a valley, he must rely on diffraction to spread the waves around the valley. The entrance to the valley is likely to be similar in size to the wavelength of the radio signals, since they can be received. The TV signals have a shorter wavelength and so diffract less into the valley, giving poorer reception.

b Reece possibly receives two TV signals – one direct and one reflected to him, possibly by buildings on a valley side. If the two signals interfere destructively his overall signal will be weak, giving a poor picture.

Comment The questions ask for a suggest-type answer. Not enough information is given to be completely specific, so use your knowledge of other situations to give a reasonable answer (i.e. your answer makes scientific sense!).

Unit 8: The Earth and beyond

1 The Solar System (page 100)

Q1 Hydrogen nuclei join together to make helium nuclei in a process known as nuclear fusion.

Comment Don't get confused with nuclear fission, which is the process used to obtain energy in nuclear power stations. Very high temperatures are required before nuclear fusion will take place.

Q2 Clouds of hydrogen gas → blue star → red supergiant → supernova.

Comment The sequence is given in the table on page 100. A supernova is only produced from a very large star. A star such as our Sun will eventually form a white dwarf.

Q3 a (i) The force due to gravity, its weight, which (ii) acts towards the centre of the Earth.

Comment Gravity always pulls towards the centre of the Earth. The satellite's weight is providing the centripetal force required for it to move in a circle.

b The satellite is always at the same position above the surface of the Earth.

Comment The ring directly above the equator at the height of geostationary orbits is full of communication satellites.

2 How did the Universe begin? (page 102)

Q1 Light waves from a star near the Earth with a lower frequency than that emitted by the star.

Comment If the emitter is moving away from the observer the frequency is reduced. In the case of light waves, a reduction in frequency results in a shift to the lower frequency (red) end of the visible spectrum.

Q2 Red shift evidence shows that all galaxies are moving apart from each other. Reversing the direction of movement of the galaxies shows that they could all have come from the same starting point.

Comment It was the famous astronomer Edwin Hubble who first observed the red shift and then proposed the Big Bang theory.

Unit 9: Particles

1 Kinetic theory of gases (page 106)

Q1 It is the lowest possible temperature, the temperature where the molecules would have zero kinetic energy (and zero speed).

Comment The kinetic theory assumes that no forces act between the particles, except during collisions. In reality, forces do occur between molecules – these would cause the gas to condense to a liquid and then to freeze to a solid at temperatures well above absolute zero. However, the idea behind absolute zero still applies – at absolute zero, the particles in the solid lattice would have zero kinetic energy and all vibration would stop.

Q2 a (i) 293 K (ii) 423 K (iii) 1273 K

b (i) 27 °C (ii) 377 °C (iii) 727 °C

Comment Remember:
temperature in °C = temperature in Kelvin – 273
temperature in Kelvin = temperature in °C + 273

Q3 1330 cm^3

Comment $\frac{p_1V_1}{T_1} = \frac{p_2V_2}{T_2}$

In this case, pressure is constant, so the equation becomes $\frac{V_1}{T_1} = \frac{V_2}{T_2}$

Remember to convert the temperatures to Kelvin: 20 °C = 293 K and −13 °C = 260 K

Rewriting the formula in terms of V_2,

$V_2 = V_1 T_2/T_1 = (1500 \times 260)/293 = 1330$ cm^3

Working to three significant figures is sufficient.

2 Developing ideas about atoms (page 108)

Q1 Similarities:

- Both contain tiny electrons.
- Both atoms are similar in size.

Differences:

- In the plum pudding model, the electrons are spread throughout the atom; in the Rutherford atom they form a cloud around the outside.
- In the plum pudding atom, the positive part of the atom is spread evenly throughout the rest of the space; in the Rutherford model, the positive part is concentrated at the centre in a nucleus.

Comment The key point is that the Rutherford model concentratres most of the matter at the centre, with most of the atom being empty space.

Q2 a Alpha particles.

Comment Gamma particles would not have deflected at all as they are uncharged. Beta particles might work, but they deflect very easily because they have such a small mass.

b Most of the atom is empty space.

c The atom must have a very high concentration of mass and positive charge. Rutherford called this the nucleus.

Comment The neutron wasn't discovered until 1932.

3 Unstable atoms (page 113)

Q1

Atom	Symbol	Number of protons	Number of neutrons	Number of electrons
Hydrogen	$^{1}_{1}H$	1	0	1
Carbon	$^{12}_{6}C$	6	6	6
Calcium	$^{40}_{20}Ca$	20	20	20
Uranium	$^{238}_{92}U$	92	146	92

Comment Remember:

number of electrons = number of neutrons in a neutral atom

number of neutrons = mass number − proton number

Q2 a 11 protons, 13 neutrons.

Comment The atomic number gives the number of protons. The difference between the mass number and the atomic number equals the number of neutrons.

b 15 hours.

Comment The count falls from 100 Bq to 50 Bq in 15 hours. It also falls from 50 Bq to 25 Bq in 15 hours.

Q3 a Beta particle.

Comment A beta particle is an electron. The 'beta' symbol can also be written as an electron, 'e'. The electron is shown with an atomic number of −1 and a mass number of 0.

b $x = 24$; $y = 12$.

Comment The mass numbers must balance on the left-hand and right-hand sides of the equation (24 = 24 + 0). The atomic numbers must balance on the left-hand and right-hand sides of the equation (11 = 12 − 1).

4 Uses and dangers of radioactivity (page 115)

Q1 a Alpha particles.

b Beta particles.

c Gamma rays.

Comment Details of these uses are given on page 114. Alpha particles are the most ionising, gamma rays are the most penetrating.

Q2 a A tracer is a radioactive isotope used in detection.

Comment Tracers are widely used to detect leaks and blockages.

b Sodium-24. The lawrencium-257 has too short a half-life; the sulphur-35 and carbon-14 have half-lives which are too long.

Comment The tracer must be radioactive for long enough for it to be detected after injection into the body but must not remain radioactive in the body for longer than necessary.

Q3 a Lead.

Comment In fact the uranium-238 decays through a chain of short-lived intermediate elements before forming lead.

b The ratio of uranium-238 to lead-207 enables the age of the rock to be determined.

Comment In a similar way the

carbon-14: carbon-12 ratio

is important in finding out the age of previously living material.

5 Fundamental particles (page 116)

Q1 For a proton (uud) the electric charge is

$$(+2/3) + (+2/3) + (-1/3) = +1$$

For a neutron (udd) the electric charge is

$$(+2/3) + (-1/3) + (-1/3) = 0$$

Comment Are other combinations of quarks possible? Yes, but to create a particle that actually exists in nature, one condition is that the electric charge always adds up to a whole number.

Q2 Suitable suggestions might be:

- Too expensive to do alone.
- Share the benefits as well as the risks.
- Expertise gained in this area may have benefits for other areas of research or technology.
- Do not want to miss out on important advances.

Comment As with all 'suggest' questions, you are looking to provide a *reasonable* suggestion. If you are going to use the idea of expense, it is usually a good idea to add a little more detail as suggested above.

6 Electron beams (page 120)

Q1 Electrons emitted from a heated cathode.

Comment The clue is in the name – 'therm' relates to heat and 'ionic' relates to electrically charged particles (electrons in this case).

Q2

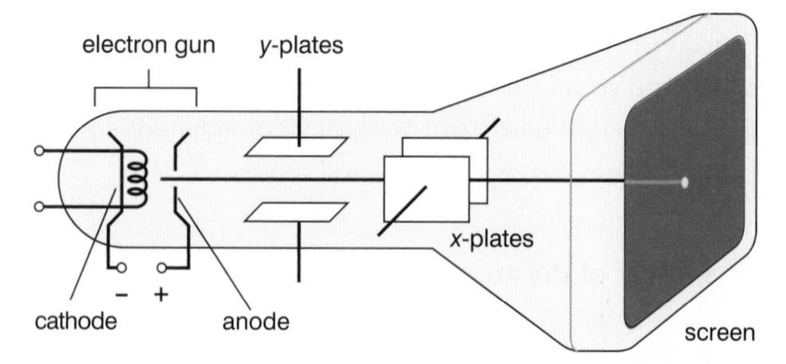

Comment Remember that *x*-plates are horizontal, *y*-plates are vertical – just like the axes of a graph!

Q3 a 0.8 V

Comment The amplitude is from the centre of the signal to the top (or the bottom). Here it is four squares at 0.2 V per square = 0.8 V.

b 50 ms = 0.05 s

Comment One complete cycle covers five squares horizontally. Each square is worth 10 ms, so 50 ms overall.

c 20 Hz

Comment Frequency = 1/time period = 1/0.05 (The time needs to be in seconds.) Remember the unit for frequency is Hertz (Hz). Look back to Unit 6 Revision Session 4 for more reminders.

UNIT 10: EXAM PRACTICE

Foundation Tier (pages 124–125)

Q1 a (i) microwaves (1).
(ii) infra red (1).
(iii) ultrasonic (1).

Examiner's comments Choosing the correct word from a list is very common at this level of question. Notice that all the waves given are part of the electromagnetic spectrum **except** for ultrasonic. You should know common uses for **all** parts of the electromagnetic spectrum and for other common waves such as sound / ultrasound.

b (i) **B** (1).
(ii) **A** (1).

Examiner's comments The wave with the largest amplitude is also the loudest and has the tallest waveform. The wave with the lowest frequency also has the lowest pitch and has the fewest waveforms on the screen.

c disturbance / vibration / movement (1).
direction (1).

Examiner's comments Describing the motion in longitudinal waves and transverse waves is tricky, so the examiners have tried to help by giving a sentence to complete and a diagram with some key words. **Always look carefully at any diagrams given.** Also remember that in a **transverse** wave the disturbance causing the wave is at **right angles** to the direction in which the wave moves.

d sound wave (1).

Examiner's comments You **must** remember that sound is an example of a longitudinal wave. Almost all other waves studied at GCSE are transverse, including all electrmagnetic waves (gamma, X-rays, ultraviolet, visible light, infrared, microwaves and radio waves).

Foundation/Higher Tier overlap (pages 128–130)

Q1 There is a maximum of 4 marks: 2 marks maximum for relevant **questions** with further marks given for relevant reasons.
Any **two questions** from:
Question relating to environment
Question relating to transport
Question relating to resource availability, e.g. fuel / water / people / space
Relevant reasons linked to the questions asked:
Effect on environment – noise / pollution / smoke / named pollutants from a power station / cables and power lines
Relating to transport – availability of road or rail links / increase in pollutants
Relating to resources – use of fossil fuels rather than renewable resources (possibly giving examples) / availability of water for cooling / availability of people as staff for the plant.

Examiner's comments The key is to give replies that are **sensible** and to relate them to particular pieces of scientific knowledge. It is easy to write answers that are too vague to score marks, such as 'Will it be good for Octon?', 'Will it be dangerous?'. Remember that this is a question asking about the wider implications of physics, so use your physics knowledge to guide you. Also, **study the diagram carefully**, there are clues to help you – Site A is near the bypass, site B is near the railway, but also close to the main street and the prevailing wind will blow any fumes over some shops and offices.

Q2 a (i) electrons (1)
(ii) negatively charged (1)

Examiner's comments This is basic recall – you need to **know** this.

b (i) One from: nucleus is positive / nucleus and alpha particle have same charge / nucleus and alpha particle are both positive (1), so they repel (1).

Examiner's comments This is a tricky question that you might not be expecting. Very often this part of the topic concentrates on the alpha particles that go **straight through** and the alpha particles that are deflected through **large angles**, whereas this question is about those that are deflected a bit. Key ideas are that the particles both carry the same (positive) charge and that like charges repel. Do not make the mistake of saying they **collide** – the diagram clearly shows that they do not – **always look carefully at the diagram**.

(ii)

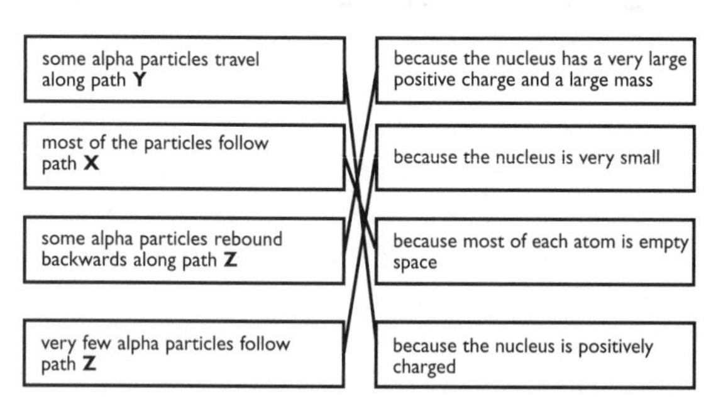

All three correct = 2 marks. Lose 1 mark for each incorrect (to a minimum score of 0).

Examiner's comments **Read through all the boxes before you start.** In this case it would be very tempting to join the top two boxes together. This would be a possible solution (and would be given a mark) but it would mean that you would end up with two lines going to one box, with another box unused. Although correct, it is more likely that, in an exam, you would feel that you had to use every box and so you would waste time trying to find a wrong answer when there wasn't one. The top and bottom boxes on the right are very similar so it is vital that you read it all though.

Higher Tier (pages 133–134)

Q1 a (i) Alternating current keeps changing direction but direct current flows in a constant direction in a circuit. (1)

Examiner's comments This is often answered very poorly, if at all. Common mistakes are to say that 'd.c. (direct current) is a steady current while a.c. (alternating current) is a changing current' – **this is wrong**. Another common mistake is to give the sources of each (e.g. d.c. from a battery, a.c. from the mains) rather than describing the **difference between them**.

(ii) Magnet (M) rotates (1), which changes the magnetic field / magnetism / field lines (1) of the iron core (1) and this causes / induces a current in the coil (C) (1).

Examiner's comments There are several places in the specification where **you need to learn a sequence of events in order**. This is one instance – the life cycle of a star is another (see the next question). If you learn these parts carefully, you will avoid the common confusions that appear in candidates' answers – it is common to see phrases such as 'magnetic current' and 'the magnet is attracted to the electricity'. **Revise carefully before the exam and you will avoid these mistakes.**

b (i) Voltage = current × resistance (1)
(ii) $V = 12 - 9 = 3\text{V}$ (1)
$R = 3/15 = 0.2\ \Omega$ (1)

Examiner's comments This question specifically asked for the equation, but **make sure you always write the equation anyway**. Here, the tricky bit is to realise that the voltage needed is 3 V. However, **if you write down your working** you will still receive the final mark even if you chose the wrong value for *V*.

Q2 Any **four** from:

Cloud of gas attracted / pulled by gravity (1)
High pressure / temperature gives nuclear fusion / reaction (1)
Energy released (1)
Expansion to red giant (1)
Collapse to white dwarf (1)
Idea of radiation pressure balanced by gravity (in main sequence star) (1)

Examiner's comments As in the previous question, here is a **sequence that must be learned**. If the points are given in the wrong order, then a maximum of three marks would be given, **so always think your answer through (perhaps writing key words in the margin) before you start to write the final answer**. Also, **read the question very carefully** – the first two words, 'our Sun', are very important here. Including these two words means that answers involving black holes will be **wrong**, because **our Sun** will not end up as a black hole – it is not massive enough. When the question appeared on an exam paper, the space under the question was left blank – **consider using diagrams where they can help your answer**. This is a good example of a situation where a flow chart type of diagram can be very helpful.

Q3 **a** force = 350 N (1) upwards (1).

Examiner's comments Total upwards force = 8000 N. When the ropes are untied, the tension in each falls to zero and so the downward force is just the weight of the balloon, 7650 N. The unbalanced force is therefore (8000 – 7650) N = 350 N. Do not forget to give the direction of this overall force.

b acceleration = force / mass (1)
= 350 N / 765 kg (1)
= 0.46 m/s^2 (1)

Examiner's comments Recall $F = ma$ and re-arrange to get $a = \ldots$ The word 'initial' is part of the question because of what follows in part (c). Do not forget the unit of acceleration.

c The resistive force increases (1) reducing the size of the unbalanced force (1), so acceleration decreases (1).
There is one communication mark for accurate spelling, punctuation and grammar (1).

Examiner's comments The resistive force is air resistance, or drag. As the balloon moves upwards it has to push air out of the way. This creates a force pushing against the balloon. The faster the balloon moves, the more quickly it has to push air out of the way, so the greater the resistive force becomes. The overall (net) force becomes smaller as it is equal to (weight of balloon – resistive force) and so the acceleration reduces. Try to think it through and give your answer in a logical sequence.

d An explanation to include: mass / downwards force decreases (1), causing an increase in acceleration (1).

Examiner's comments Throwing the bags overboard reduces the weight of the balloon – it becomes lighter. The upward force does not change (because that is the lift provided by the air in the balloon), so the overall forces become more unbalanced – there is a larger overall upwards force, causing a larger acceleration. Remember to include the comment about the acceleration increasing – that is what the question asked for. **Remember to check that you have actually answered the question asked.**

INDEX

William Collins' dream of knowledge for all began with the publication of his first book in 1819. A self-educated mill worker, he not only enriched millions of lives, but also founded a flourishing publishing house. Today, staying true to this spirit, Collins books are packed with inspiration, innovation and practical expertise. They place you at the centre of a world of possibility and give you exactly what you need to explore it.

Collins. Do more.

Published by Collins
An imprint of HarperCollins*Publishers*
77-85 Fulham Palace Road
London W6 8JB

First published 2005

10 9 8 7 6 5 4 3

ISBN-13 978-0-00-719056-0
ISBN-10 0-00-719056-5

Malcolm Bradley and Chris Sunley assert their moral rights to be identified as the authors of this work.

British Library Cataloguing in Publication Data
A catalogue record for this publication is available from the British Library.

Acknowledgements
The Authors and Publisher are grateful to the following for permission to reproduce copyright material:
AQA: pp. 128, 129, 131, 132,
Edexcel: p. 126, 127, 134 (bottom)
OCR: pp. 122, 123, 124, 125, 130, 133, 134 (top)

Photographs
(T = Top, B = Bottom):
John Birdsall Photography 11T; Andrew Lambert 8, 11B
Science Photo Library 78, 86.

Illustrations
Roger Bastow, Harvey Collins, Richard Deverell, Jerry Fowler, Gecko Ltd, Ian Law, Mike Parsons, Dave Poole, Chris Rothero and Tony Warne

Every effort has been made to contact the holders of copyright material, but if any have been inadvertently overlooked, the Publishers will be pleased to make the necessary arrangements at the first opportunity.

Edited by Margaret Shepherd
Series and book design by Sally Boothroyd
Index compiled by Ann Lloyd Griffiths
Production by Katie Butler
Printed and bound by Martins the Printers Ltd, Berwick upon Tweed